The All-American Boys

The All-American Boys

Walter Cunningham

with assistance by Mickey Herskowitz

Macmillan Publishing Co., Inc.

NEW YORK

Macmillan Publishing Co., Inc.
866 Third Avenue, New York, N.Y. 10022
Collier Macmillan Canada, Ltd.

Library of Congress Cataloging in Publication Data

Cunningham, Walter, 1932–
    The all-American boys.

    1. United States. National Aeronautics and Space Administration.   2. Astronauts—United States.
I. Herskowitz, Mickey, joint author.   II. Title.
TL521.312.C86   387.8   77–22721
ISBN 0–02–529240–4

FIRST PRINTING 1977

Printed in the United States of America

To LO, BRIAN, *and* KIMBERLY *for coping*
so beautifully during the period I
was having the time of my life and for
their patience during the four years
it took to write about it.

Contents

Preface

W R I T I N G A B O O K is an unusual experience. It makes people nervous, especially those being written about. Shortly after publication of the first astronaut-authored book I received a call from the Astronaut Office. I could almost hear the caller's pulse throbbing at the other end of the line. "We have a copy in the office, but the boys are half afraid to open it. I hear it tells *everything*. I sure hope your book isn't going to get into a lot of that gamey stuff."

As I hung up it occurred to me that the world might not be ready for "everything."

Why another book on the astronauts? Well, why not? Why three volumes on the fall of the Roman Empire? Why a sequel to the *Happy Hooker*?

In the sixties NASA set out to fashion its image, but the myth of the super-hero astronaut was purely a creation of the news media. Most of us found it flattering and easy to go along with. Some even cultivated that image, but few could measure up to it. Most of us recognized it was unlivable only slightly before we realized we were stuck with it for the rest of our lives. We will remain trapped in that image until the public takes off its rose-colored glasses and begins to see us as people. In this book I have attempted to strip away the veneer and tell how America's most famous heroes were made; and why; and what happens to them. We weren't all simon-pure nor all hell-raisers. Some of history's greatest deeds were accomplished, and a remarkable adventure was fulfilled, by men who were all too human in their weaknesses as well as their strengths.

We have been in the public eye for seventeen years, but how we think and work, act and react has never been told. Here is the inside story of NASA and our first fifteen years in space and of the principal actors and stars of that experience—the astronauts. My objective is to share the enthusiasm and skill we brought to our work as well as to tell about the warts and moles which sometimes compromised it. The shortsighted goal of "a lunar landing in this decade" was a disservice to an undertaking which will be a permanent fixture in our future. It was, and is, a whole lot more than *just* a lunar landing. It isn't yet history; it has only just begun.

This book is not a definitive account of NASA or its people. And it is not an autobiography of Walter Cunningham, although it is bound to reveal as much about myself from what isn't said as from what is. It is but one man's perspective. It began in 1961 in the curve of a mountain road above Los Angeles.

This supreme adventure is only now moving into high gear. It is no more possible to assess its significance today than it would have been in 1515 to place a value on Columbus' discovery of the New World.

The All-American Boys

1

"We've Got a Fire in the Cockpit!"

FROM THE ROOF of the building where the flight crews lived, slept, and maintained offices, we could see the skeletal outlines of the gantries, standing like great oil derricks on a stretch of sand along the Atlantic at Cape Kennedy. Across the palmetto trees and slash pines, beyond the ribbon of blue water that circled Merritt Island, it was a beautiful sight.

Some days we would sit up there, peel off our shirts to catch the sun, and study the flight plan or mission rules. But this was not one of those days. It was the Friday of a long week, and across that eight miles of Florida jungle, on Pad 34, they were testing the Apollo 1 spacecraft for the first step in a series that would fulfill man's dream of walking on the moon.

The backup crew—Wally Schirra, Donn Eisele, and myself—had its own drills that day. Our presence was not required at the pad, but we were sweating out the test and clock-watching anyway. We planned to fly home to Houston and visit our families, but it would be bad form to leave before the prime crew had finished its chores: who knows, we might be needed.

Our offices were on the third floor, just down the hall from the control room for all the spacecraft tests. Every half hour or so one of us would wander in and ask how things were going.

They were going slowly, and the usual cool professionalism was now and then giving way to emotion. At one point the intercom between the spacecraft and the blockhouse began to act up, static drowned out any voice contact. "How the hell can we get to the moon," snapped Gus Grissom from the Apollo craft, "if we can't talk between two buildings?"

Grissom and his crew, Ed White and Roger Chaffee, were performing the ship's first full "plugs-out" test. Anything that could be disconnected, was disconnected, and the Command Module was left to run on its own internal power. It was not a complicated test and was labeled, at that time, nonhazardous because there was no fuel on board. There had been some discussion about whether to run it with the hatch open or closed. The hatch was a brute of a thing that took at least ninety seconds to open and required help from the outside if it was to be done without damaging it. They decided to close it.

Only the night before our backup crew had run the plugs-in version of the same test, with auxiliary power cables connected to the spacecraft through the open hatch. For Apollo 1 it had been routine, and we finished around 8:00 P.M. After being debriefed, we showered, changed, ate, and drove into town—Cocoa Beach—to raise a little hell.

The test that Friday—January 27, 1967, almost one month to the day before the launch—was to begin at 11:00 A.M. That would assure all of us of an early takeoff for Houston. But from the outset there were nagging delays. It was after 1:00 P.M. when Gus, Ed, and Roger finally maneuvered into the cabin, awkward in their Buck Rogers space suits. All over the area, the chores of the day came to a halt as they rode the elevator to a height equal to twenty-two stories and walked across the catwalk to the sterilized "white room." At such moments workmen often broke into applause. Every eye was on them. I was struck again by the ease with which the spotlight shifts to the crew making the next flight, the naked admiration of the technicians who followed their every step. It was once again the Hour of the Astronaut at the Cape. Show time for Apollo 1.

There's no escaping the feeling of envy when you're on the backup crew.* The backup crew blends into the scenery and does it willingly, gladly, because in the recesses of each one's mind a voice is whispering, "My turn will come."

The crew took their places and were strapped into their couches. Once they stopped the countdown when Gus complained of a sour odor in the cabin. The test conductors considered scrubbing the test but then decided to press ahead.

Wally, Donn, and I kept checking our watches. The day was wearing

---

* Wally, Donn, and I had been in that support role for ten weeks, since the cancellation of our own mission.

on. By 5:00 P.M. one of us said, "Hell, those guys might be out there all night. Let's bug out." We had to be back at the Cape in forty-eight hours. It was a simple test, a piece of cake. What could go wrong? The prime crew would probably finish late and fly to Houston the next day.

We jumped into three separate cars for the half-hour race to Patrick Air Force Base. We traveled light—flight suits and shaving kits. At Patrick we strapped on a pair of T-38's. I was at the controls of one trainer, with Wally in the rear cockpit, and Donn flying formation in the second. With clear skies and no headwind, we enjoyed a loose, easy flight home, touching down at Ellington Air Force Base at dusk. We were in good humor, lighthearted, with that Friday out-of-school feeling.

Then we saw Bud Ream.

We had taxied up to the NASA hangar, where we were usually met by the lineman who begins to get the airplane ready for the next flight. That night he was joined by Bud Ream, the Number 2 man in flight operations. Bud was a NASA test pilot, short, solid, with a crisp manner. There was no reason for him to be there. He stood on the concrete apron, waiting for us. His face was grim.

He motioned for us to follow him up the stairs to his office. As we walked he spoke, his voice hoarse and tense, "There's been an accident at the Cape. The prime crew has had an accident."

An accident? How could they have had an accident? They were on the pad. A fall on the gantry? A car accident? At such moments, your mind spins. Downstairs my wife, Lo, was waiting in our car with the kids. They had watched us taxi in, and she would be wondering what was taking so long. All we usually did was drop off our flight equipment and head out.

At first no one spoke. We just looked at each other. Then the questions began pouring out as we all talked at once. What accident? Who was hurt? How? When?

"The crew is dead," Bud said. "They're dead. All of them."

Suddenly, Lo appeared at the door. She had seen Bud walk out to meet us and soon after had heard a news bulletin on the car radio. Something had gone wrong at the Cape. No details.

Bud Ream was making an effort to keep himself under control. He had been close to Gus ever since 1959, when Project Mercury, our first manned program, began. He also knew Roger and Ed well. He was shaken. We all were.

Someone got on the phone and dialed Mission Control at the Johnson Space Center, five miles away. The men of the NASA command who were in Houston at the time had already assembled there. They were replaying the tapes of the test, trying to determine what had happened.

We spoke with two of the brass, George Low and Chris Kraft. They gave us all they had: There had been a fire. It had flashed through the cabin in seconds. They were burned alive.

Whatever illusions any of us had had about our jobs were temporarily forgotten that night. Shock had replaced envy. If we had ever talked about death—which we had not—none of us could have imagined it coming as it did: to three men, helpless, in maybe half a minute, only 218 feet above the ground.

And I remember one of us said, "Thank God, it happened on the pad."

That may sound like a cold and strange way to think. But in a time of crisis or tragedy or embarrassment, a convenient process takes hold of men who live by flying. They go on instruments; instincts take over. The crew was dead; nothing could be done for them. The important thing now was to find out how and why—to protect the living. It was that simple. Even as we felt the first dull shock at the deaths of Gus Grissom, Ed White, and Roger Chaffee, we thought, "What about the program? What about us?"

I doubt whether we were very different from the rest of the country. We felt the effort would continue, that America still had to reach for the moon. But it was obvious that there would be no more manned flights until we knew what had gone wrong with the spacecraft that had been designed to get us there.

But at least the evidence was right in front of us—on Pad 34.

That was important, not just for scientific or technical or even humane reasons, but for our own sense of order. It meant that we had a chance to discover the cause of the fire before another craft was wheeled out of the barn.

You have to understand the double-edged sword that hangs over every aviator's head. Think of the mental conflict: your buddy has just been killed in your favorite fighter plane. You respected him as a flier. You want to believe that he did everything possible to stay alive—that he was undone by some perverse fate that caused an obscure part to fail. But a shadow crosses your mind. If that's the case, then it could happen to you; it could be *your* body going home in a box with a flag draped across it.

So when the board of investigation brings in a finding of "pilot error" you breathe easier; the equipment didn't fail. It's better to believe that the other fellow blew it: "Poor guy, he bought the farm. But it can't happen to me. I'm too smart, too good, too cool."

Of the eight astronauts who have died, four were in air crashes: Ted Freeman hit a goose, Elliott See and Charlie Bassett hit a building in St. Louis, and C. C. Williams hit the ground in Florida. In none of these cases did the accident board attribute the cause to pilot error, but in each the deaths could have been laid to the pilot's hand. In truth, most pilots kill themselves.

But not this time. And so we thanked Providence that we could look for answers. All along our fear had been that the hairy mishap, when it came, would occur in outer space, leaving no traces, offering no clues—only an eternal silence.

The truth of this is not easily understood. But the death of Gus Grissom's crew at the Cape made it possible to land a man on the moon on schedule. Indeed, it may have saved America's space program. So we cannot consider their deaths to have been in vain.

The fire did something else. It reminded the American public that men could and would die in our efforts to explore the heavens. It re-created the tension and uncertainty that had greeted the early flights of Shepard, Grissom, and Glenn. We had reached a point where the public was aware of space only when a crew was out in it. That environment, they knew, was strange and dangerous. The fire on the pad made it obvious that even the daily routine of the astronaut involved a risk. It once more placed us, in the minds of the public, in that category of men who roll the dice, who—like race car drivers or bullfighters or the cliff divers of Acapulco—put their hides on the line every day.

If we didn't see ourselves that way, it didn't make it any less so. It wasn't so much that we accepted the risks, but rather that we had never admitted the risks existed in the first place. It was all tied up in what the writer Tom Wolfe once described as "The Brotherhood of the Right Stuff." It is part of what sets flying apart as an art, as a science, as a way of living. And dying!

Wolfe came as close to describing it as anyone ever has: "The main thing to know about an astronaut, if you want to understand his psychology, is not that he's going into space but that he is a flyer and has been in that game for 15 or 20 years. It's like a huge and very complex pyramid, miles high, and the idea is to prove at every foot of the way up that pyramid that you are one of the elected and anointed ones who have *the right stuff*. . . .

"The right stuff is not bravery in the simple sense; it is bravery in the most sophisticated sense. Any fool can put his hide on the line and throw his life away in the process. The idea is to be able to put your hide on the line—and then to have the moxie, the reflexes, the talent, the experience, to pull it back in at the last yawning moment—and then to be able to go out again the next day and do it all over again—and, in its best expression, to be able to do it in some cause, in some calling that means something."

Whatever it was that qualified us to be astronauts—whatever the chemistry that created in us the *need* to fly—also gave us a different view of life and its challenges, of the earth and its treasures. Our response at any given moment was not likely to be conventional. From the outset, the space program was related through spiritual bloodlines to the era of scarves and goggles, to Lindbergh and Rickenbacker and the Red Baron, and to all those old Chester Morris movies and the immortal line, "My gawd, they're not going to send the kid up in a crate like that?"

If you have been a jet fighter pilot for any length of time, you have seen your friends get killed—often. You build up a certain immunity. I flew with such men and knew them well, men whose faces I can no longer remember. They are frozen in time now as shadows in old group photos. My younger brother Ken will always be the twenty-nine-year-old fighter pilot.

Two weeks before I was to report as part of the third generation of astronauts, Ken was killed flying an F-104 at Webb Air Force Base in Midland, Texas. We had grown closer as we grew older. Our interests, our horizons, were the same. Ken had been the only member of my family who could fully appreciate and share the enormity of the space program and what it meant to me. Yet when I heard of his death my emotions remained intact. I was conscious of doing what was expected of me under the circumstances.

That weekend I had flown with my marine reserve unit, after which Lo and I went to visit my parents. We walked into their home to find everyone crying. They had just received the telephone call notifying them that Ken's plane had crashed. It was Sunday, December 15, 1963, ten days before Christmas. He left behind a pregnant wife and a small daughter.

Quickly, I called the base. Talking to the duty officer, I learned that Ken had crashed while returning from an air defense exercise. When he had activated his landing flaps they had lowered only on one side. The plane had spun, almost like a top, and corkscrewed into the earth. One of his closest buddies had been killed two weeks before in a similar accident.

Ken had been an excellent aviator, and as I flew to Midland that night I thought how fortunate he had been to spend seven years doing what mattered most to him.

My mother, of course, felt flying had killed him.

If you are attuned to flying, it can change you. It remakes you. Your heart beats stronger and your ego grows larger and your instincts become more sensitive. Life takes on a dimension that people whose feet are anchored to the ground can never know. I have rarely strapped myself into the cockpit of a jet airplane without consciously thinking how lucky I was.

Flying is a death-oriented business. You either accept the odds or you stay the hell out. No one had understood this any better than Gus Grissom. "If we die," Gus had once said, "we want people to accept it. We hope that if anything happens to us it will not delay the program. The conquest of space is worth the risk of life."

Prophetic? A portent of the tragedy to come? I don't think so. Gus was just expressing what we all felt. No other attitude would have been tolerable.

When a friend or even a brother dies in the crash of his jet, we tend to grieve less for him than for those who couldn't understand why a man would be willing to die for something. We believe there are worse things than dying. Maybe it sounds like happy nonsense. Or a pose. Or superstition. But that was the world in which we chose to live. None of us in this trade comes to death unprepared. Yet it wasn't something we thought or talked about, not now, and not on the night of January 27, 1967.

Yet it was hard to accept the fire. We had run virtually the same test the night before, with the continuing irritation of "glitches," those minor problems we had almost come to expect. We kept telling ourselves, and so did the engineers, that we couldn't expect the first article out of the factory to be bugproof. Wally had complained that the same problems kept showing up. In the push to keep the schedule moving they were not being effectively worked out. His words to Gus were, "If you have a problem, even a comm [communications] problem, get out of the cabin until they've cleaned it up." Wally was always more inclined to say, "Hey, it's not working. Call me when it's right."

We thought of that later. We thought of the weeks and months of griping about the test failures and the marginal workmanship that kept surfacing as the craft was being built. Now it came back to bite us. We cursed the spacecraft that killed them.

During one test the nozzle of the spacecraft engine had shattered. During another the heat shield split wide open and the Command Module sank like a stone dropped at high speed into a water tank. There had been fuel ruptures. The cooling system had failed. It had been one damn thing after another. Now three men were dead.

I tried to imagine what it must have been like for those at the Cape who stood helpless and listened to them die. "It was about twenty minutes," Joe Shea, the Apollo program manager, said, "before we realized that they were not about to get out. It was pretty tough."

Wally, Donn, and I were still in our flight clothes. We stayed in them for the next six hours. We stopped at the Johnson Space Center, in

Clear Lake, to find out what else we could learn. It was all shock and confusion and consternation. Then we did what pilots usually do at such times. We fanned out to visit the wives.

My children—Brian, age seven, and Kimberly, age five—were shaken up but not quite sure why. They had sensed this mood before. Their playmates included the children of Ted Freeman, Elliott See, and Charlie Bassett—all of whom had died in the preceding thirty months. They had begun to notice that some of Daddy's friends sometimes didn't come home from work.

After we dropped the kids at a neighbor's house, Lo and I headed for the home of Martha Chaffee. The news media had already collected in the front yard, but the neighbors, most of them involved in the space program, had rallied around to protect the families.

Lo is a very pretty woman with brown hair and bright eyes. She has spirit and an active mind. But having gotten married one month before my release from active duty, she had little experience in the rituals of a fighter pilot's wife—such as comforting the widow. Now here she was, nervous and uncomfortable, standing with me at the front door of the first astronaut family we had met, only three years before.

Inside was the usual mourning scene. Those doing the comforting seemed more uncomfortable than the comforted. The role of the bereaved is at least spontaneous. In this case, the look and kitchen conversation of each wife left little doubt that they felt, "There but for the grace of God. . . ." The husbands were brought face to face with something seldom discussed and generally ignored: the fear our wives felt about the jobs we loved. The guys saw a lot more of each other than the wives did, but we maintained a more impersonal relationship. It was much easier for the wives to empathize with each other.

Lo was closer to Martha than to any of the other wives and knew the preceding year had not been a smooth one—for any of the widows or any of the wives. The system worked against them. Training kept their husbands not only occupied but preoccupied as well.

Martha was still glassy-eyed and had obviously not yet come to believe that Roger had really died.

With the sudden cancellation of our own flight on the second Apollo mission, in one week we went from testing spacecraft 014 to performing similar tests on Grissom's spacecraft 012. Even our training picked up for a while because of the higher priority on the use of the mission simulators. Making the best of what we accepted as a bad deal, I was pleasantly surprised. I found Gus to be a decisive guy, a team leader, and an independent thinker, who nevertheless encouraged input from the rest of the crew. It was a nice change to work with him.

Gus, a taciturn, grizzled fellow, was scheduled to become the first man to make three space flights. An air force cadet at eighteen, he had flown a hundred combat missions in Korea. One of the stories told about him was how, when he first got to Korea, he found that pilots who had not been shot at by a MIG weren't allowed a seat on the bus to the hangar. Gus stood only once. He had shot it out on his first mission to qualify for a seat—and the "brotherhood of the right stuff."

Gus had backed up Shepard's historic first Mercury flight. Five weeks later, aboard Liberty Bell 7, he had become the second American in space. When his spacecraft hatch exploded off after splashdown and the capsule flooded, Gus swam to safety and lived to be periodically embarrassed as the only man ever to lose his spacecraft.

He filled the bill as the prototypical test pilot. He could sit at a bar for hours, and he never failed to notice a pretty girl in the room. He was into the race car scene with Gordon Cooper and Jim Rathman, the Indy driver. He could be cranky and tough, but he went his own way. He wasn't a hanger-on.

Betty Grissom was the only one who didn't cry. She was stoic, composed. The Schirras, who lived next door, were quickly on the scene, and Wally was trying to take care of things. But Betty was a veteran. She had traveled a lot of miles with a much rougher guy than had the other wives. She had obviously made her own adjustment to that fact, kept her calm and her dignity, and from that night on she fought for Gus Grissom's memory. Gus had been a professional and believed what Betty soon forgot: if you lived by the sword, you could expect to die by the sword.

Some months later Betty was invited to the White House to accept a Distinguished Flying Cross awarded posthumously to her husband. She declined. "Medals don't mean much to me now," she said.

Ed White was one of the second generation astronauts. On June 3, 1965 he became the first American to walk in space and "felt red, white, and blue all over," as he described it later. He was, I think, consciously cultivating the All-American–John Glenn image. That was Ed—at least most of the time. The rest of the time Ed White was not much different from the rest of the guys. Take the way he ran the Astronaut Training Gym. It originally contained one handball court and one squash court. Ed and two or three others were the only squash players in the group; the rest of us were dedicated handball addicts. When money was budgeted for two more courts Ed began to plan for one more of each. When we were polled as to our preference for two more handball courts or one of each, the vote came out something like seven to one for the handball courts. Running into Ed at the gym one day, I asked, "Did you see the poll?"

Ed's reply was short and to the point: "Yeah, I've seen it, but as long as I have anything to do with the project, there is going to be one more of each." Ed was right, and to this day the guys wait in line to use the two handball courts while the two squash courts generally stand empty.

The son of a retired air force general, Ed was a West Point graduate. He was also an athlete who barely missed qualifying for the 1952 Olympics as a 400-meter hurdler. Ed was a golden boy, meticulous, tall, clean-cut, and a fierce advocate of all the basic virtues, God, country, Mother, and religion. On Gemini 4 he took along a gold cross, a St. Christopher medal, and a Star of David.

The grief of his wife Pat was visible—she took it hard. But she also felt the sense of public loss; she had shared her man with the nation. Theirs had been an All-American marriage, a tighter one than most of those buffeted by the winds of space. Ed wasn't the kind of guy who parked at bars, drinking and flirting—the more popular fighter-pilot recreations.

Roger Chaffee was the rookie of the crew and one of my contemporaries. He was one of the younger astronauts, with a short but illustrious navy career behind him. He had flown most of the photo missions over Cuba, mapping the missile sites during the "eyeball-to-eyeball" 1962 missile crisis.

Roger was success-oriented and had his eye on the moon landing. He and Martha had had a lot going for them; now it was gone.

Out of all the despair and remorse and uncertainty, we were sure of one thing: there would be hell to pay before another spacecraft was launched. Wally and Donn and I figured it would be months before our presence was required again at the Cape.

Saturday, the day after the accident, doctors at the Cape asked for a uniform in which to bury Gus, and Wally asked if I would fly it to Patrick Air Force Base. That afternoon I quietly pulled my Porsche into Wally's drive, slipped next door, and took the uniform from Wally, who passed it through a partly opened door. He had sneaked it out of the bedroom closet to avoid upsetting Betty.

When I landed at Patrick one of the doctors was waiting at the hangar for the uniform. I recall thinking that the flight was a little unusual. Normally, flight surgeons attempt to keep aviators grounded after the loss of close associates, because they might become distracted and thereby contribute to another accident. Whether it was the non-flight nature of the accident or whether they considered us more stable, I was glad to have something to do that moved things back toward business as usual—anything that didn't bring me too close to the survivors' grief.

Memorial services for the three astronauts were held in the Clear Lake area that weekend. I was in the flyby for the ceremony for Ed White—the "missing man" formation.* A quote from Ed's pastor, the Rev. Connie Winborn, appeared in *Time* magazine: "Let us not expect to sing the victor's song, unless we are willing to risk the harsh notes of tragic loss and personal sacrifice."

I dread funerals. I don't like to get that personal with death. But we all flew to Washington to attend two funerals at Arlington National Cemetery the following Tuesday: Gus' in the morning and Roger's in the afternoon. Ed White was buried at West Point.

Donn and I left our wives in Washington and flew directly to the Cape by T-38. We intended to clean out our quarters and retrieve our clothes.

I finally made it home three weeks later. We were both put to work with the official Board of Inquiry. I was appointed to the Emergency Provisions Review Panel, one of twenty-one established to investigate every aspect of the catastrophe. Like everyone else, I was aware that if we couldn't come up with some right answers, the program might not survive. We all knew that the climate was ripe for an emotional backlash against the space effort. The accident would have been even more tragic if the crew had died for nothing and the work they started had not been allowed to continue.

Donn and I were quickly caught up in the state of urgency that gripped the Cape Kennedy complex. Wally remained in Houston, still smoldering over the deaths and railing against the inadequacy of a dumb machine built by contractors who should have known better. He was attending to Gus' personal affairs and adjusting his thinking to the long delay we all knew was ahead.

Some of the more ghoulish details had been completed by now. The offices of the dead astros at the Cape and back in Houston had been sealed to prevent the curious from stealing souvenirs. Certain effects had been removed from the spacecraft but their bodies had remained in place, in the charred cockpit, for seven hours before being brought down. No one wanted to risk disturbing the evidence. And no one really knew what the hell to do. There were simply no procedures. The United States, after all, had never had any of their space heroes die ignominiously in a fire on the ground.

The investigation was already on a twenty-four-hour-a-day crash basis. That first night Frank Borman, Donn, and I found an empty studio in one of the small control centers and began playing the last

* A flight of four with one spot vacant. It is the aviation expression of the riderless horse in an army funeral procession.

voice tapes from the spacecraft. Only a handful of people, other than the controllers in the blockhouse at the time of the accident, had actually heard them.

We listened over and over, trying to be detached yet mesmerized by the few seconds of sheer, stark terror that emerged from the tapes. I had a sick feeling at the pit of my guts—not just because someone had died, not just because they had been friends and companions, but because of the horrifying way they had to die.

Fire! It is the most detested of a pilot's enemies, slower and more painful than any other—and sometimes not sure enough. Yet the possibility of a fire on the ground had been given precious little thought. The cabin didn't even have an extinguisher. The Apollo 1 spacecraft was a bomb fused with a 100 percent-pure oxygen environment, the most inflammable kind of atmosphere.

As we concentrated on the crucial last half-minute, we had two objectives: determine who was saying what and—if we could—to reconstruct what the crew was doing during that time. We hoped desperately to find a clue to the problem they had encountered in the last minute of their lives.

It was an ordeal to listen to the tapes, starting, stopping, backward and forward, trying to isolate each split-second. In the end, a lot of people came up with different versions of what was said. The engineers at North American Rockwell had their version. So did Bell Laboratories, where the voices were processed by computers, analyzed, and broken down into components. There has never been a unanimous opinion on what was said or by whom.

According to the tapes, the first indication of the fire came at 4.7 seconds after 6:31 P.M. with the crew's announcement of alarm. Fourteen seconds later the inner shell of the cabin exploded. It is unlikely that they could have lived beyond 6:31:20. As I interpreted the tapes, all the transmissions were from Chaffee. White was working to remove the hatch and Gus was probably trying to help.

"Fire!"

"We've got a fire in the cockpit!"

"We've got a bad fire. . . . Let's get out. We're burning up. . . ."

The only other transmission was a very brief scream. Total elapsed time: twelve seconds.

We went back to the crew quarters that night and had more than one stiff belt. The end had come quickly, but what a horrible way to die!

The blackened, burned-out capsule sat in the "white room" under very close control. Every last bit of ash was sifted, a false floor was suspended in the cockpit and a thousand other precautions were taken to keep from disturbing any possible evidence and still the cause was never determined.

The investigation's conclusions became a virtual indictment of everyone connected with Apollo, including those of us who were to fly it. We contributed to the disaster by our willingness to tolerate questionable designs, equipment, and testing procedures; by ignoring our own good sense and accepting borderline safety margins; in short, by our eagerness to blow the bolts and get off the ground.

All of us who carried out the testing at North American Rockwell felt that some of the systems were inadequate and that crew safety had a lower priority than in the past. NASA program management and contractors alike were marching to a different drummer than the flight crew on this one. The birthing pains of Mercury and the just-completed Gemini program seemed like perfection alongside Apollo.

Many things had needed changing, but changes meant delays. The program manager had other things to weigh besides the crew's demands for performance and safety. They had to make daily trade-offs between budgets, schedules, the payload of the spacecraft, and operational desires.

But while we had been alarmed by all this, we had no premonition of what was to come. The closer we had gotten to launch date, the more omnipotent we had felt. We had already been looking ahead to the next spacecraft, 101, into which would go some of the improvements that we had been unable to obtain for Apollo 1. All subsequent Apollo Command Modules would be Cadillacs by comparison, designated Block II spacecraft. Spacecraft 101 was the first of the new line.

There had been so many changes to the original design that the first two Block I spacecraft, 012 and 014, were the only two produced.

OK, we had shrugged, so spacecraft 012 was a piece of junk. Any astro worth his salt would fly the crate anyway—or die trying. This was no time for the right stuff to waver. The simulators and mockups had been well behind schedule. From our vantage point as the backup crew, Wally, Donn, and I had seriously doubted that the prime crew would be sufficiently trained to meet the launch schedule. And certainly none of *us* had been trained well enough to step in, if it became necessary. But ready or not, the program went forward. There had been pressure within NASA to get off on time. In a very real sense time was money and the annual funding battle with Congress had been getting tougher.

No one had wanted to sound the alarm, to be caught holding the umbrella. North American, the prime contractor, wasn't about to say, "Whoa, we need more testing, hold the flight." And our boss, Deke

Slayton, wasn't going to be the guy who told NASA, "Hey, we're not going to be trained in time"—Deke's boys were *always* ready. The bugs were still being worked out of the simulators, but the flight crew wasn't going to raise the umbrella, either.

So many things weren't working on the Apollo simulator that one morning Gus had hung a lemon outside the entrance hatch. But their attitude—and we all shared it—had simply been, "Blow the bolts and we'll do whatever the hell is necessary after we get in the air."

As the fire investigation proceeded, it at least became clear that Washington was not going to kill the program. Our main worry was that, once the Board of Inquiry made its report, there would be a tendency for the space agency to overreact, to load up on so much safety equipment we wouldn't be able to get the damn thing off the ground. We wanted a better spacecraft, but we wanted it while we were still young enough to enjoy it. We didn't want one so safe that it wouldn't fly. We couldn't reach the moon in an armored tank.

Frank Borman had been assigned to the Board of Inquiry and represented the flight crew—and himself—admirably. He handled himself with professional assurance and provided credibility to an investigation that many feared would be a whitewash. Frank is a practical and extremely able guy. As you read the names along the halls of astronaut offices, you see so much rank—all those colonels and navy captains—that it looks like the Mexican army. But Borman is one of those who would have reached that elevated status regardless of what career he chose. The same can't be said of all the others.

My involvement was finished in a matter of weeks, but the inquiry itself went on for nearly a year, with the first reports submitted in April, three months after the fire. Experts had microscopically examined the cockpit, the blobs of melted wires and the scorched dials and toggle switches of the instrument panel. A partially completed Apollo spacecraft—the very one Wally, Donn, and I had trained to fly—was flown to the Cape and cannibalized so that its components could be compared piece by piece to the blackened debris scattered about the ruined craft.

The board was never clearly able to establish the cause of the fire but indicated a short circuit in the insulated wiring beneath the left-hand couch, which Grissom occupied, as the most probable cause. The wiring itself was found to be defective.

The exit hatch, of course, was a monster, heavy as hell, awkwardly placed above and *behind* the middle seat. It had to be cranked to get it open. A fellow like Ed White, strong as a young bull, had to reach back over his head and strain for minutes to release it. It was apparent from the position in which he was found on the couch that he had made a desperate stab at just that.

Escape provisions, rescue and medical assistance plans, wiring, and plumbing were all found to be inadequate. There were deficiencies in design, manufacture, quality control, and testing.

Too little attention had been given to crew safety. The pure oxygen atmosphere, an unwarranted amount of combustible materials in the cabin, careless workmanship, and poor judgment—all had been contributors to the tragedy. Proper emergency procedures had not been established. Fire, rescue, and medical teams had not been present since the test had not been labeled hazardous.

In short, Gus, Ed, and Roger never had a chance.

The indictments went on and on. The investigation also established that they died of asphyxiation. The inside temperature reached more than 1,000° Fahrenheit.

Step by painful step we began putting the pieces together.

NASA shifted personnel, clamped down hard on test procedures, and put a foot to the neck of North American. At a cost of $75 million, the Command Module was redesigned. Among the changes was a new one-piece hatch that swung out and could be opened in ten seconds.

To snuff out any fire that might develop, an emergency venting system was added that could reduce cabin pressure in seconds. And while the craft was on the pad, a mixture of 60 percent oxygen and 40 percent nitrogen was substituted for the 100 percent oxygen formerly used.

From the astros' point of view, the changes encouraged the belief that a spacecraft would be built that could, at last, perform its intended mission. The relentless pressure to meet a schedule was gone for the first time since the program's inception. With the federal purse a little looser, to make the spaceship safe for "our boys," many changes we had campaigned for, only to be turned down, were suddenly alive again. A rapid repressurization system, an emergency oxygen breathing system, electric power system changes—from brand-new crew couches to an additional urine dump nozzle, there was scarcely a system that didn't benefit from the exercise.

And out of the whole mess, North American was to bring forth one of the greatest machines ever built by man. I am convinced that it would not have been possible to reach the moon in only five missions had we not gone through this rebuilding process, which was the inescapable result of the fire on Pad 34.

Within three weeks of the tragedy those in our office were already shedding their guarded, solemn attitudes and guessing who would get the next flight. That might strike some as dancing on the coffin. Here we were, still laying to rest the facts of that nightmarish accident, and we're wondering who would succeed them. It was an echo of the days of the old Roman gladiators, when the crowds would hail the new champion even as the old champion, now dead, was being dragged from the arena.

But the program was going on. Someone would fly the next Apollo mission, a crew still had to test the craft in space, and by God I wanted to be on it. The closeness of my position to the crew of Apollo 1 only strengthened that resolve. I felt it would be a gratuitous low blow if our crew didn't fly the next mission.

It was five weeks after the fire when Wally, Donn, and I flew into the Cape for a brief visit. We had finished dinner one night and the three of us were sitting in the living room of the crew quarters. Deke Slayton walked in. We passed a few amenities before Deke casually said, "I want to let you guys know that you have the next flight."

I looked quickly at Donn and Wally. We nodded. There was no show of emotion. Decorum required that a certain amount of regret be shown for the manner in which the assignment was inherited.

By the time I completed my work on the inquiry and finally left the Cape to return to Houston, it was impossible to contain my feelings. As I drove toward Cocoa Beach, I crossed the bridge over the Banana River. Instinctively, I looked toward Pads 34 and 37, out at Moon Base, USA, with all those gantries and launch towers lined up along the beach like toys in a giant Erector set. Pad 34 was now empty again, the charred spaceship removed at last.

And I thought, "By God, the next time a man lifts-off from here, it will be me."

The hair on my arms stood up. I was a step closer to flying in space and I luxuriated in that feeling all the way home.

2

You Lucky Sonofabitch

I HAD SWUNG my Porsche off the road and parked it in a bend at the top of the Santa Monica Mountains. The city of Los Angeles sprawled far below, a hazy, sweeping panorama. Beyond it stretched the California coastline, miles of sand and surf and tacoburger huts.

It was a picture-postcard view, but at that moment I didn't see it. I was concentrating on my car radio, oblivious to everything else. My hair could have been on fire and I wouldn't have noticed.

My trance was broken by a rush of words exploding in my ears, shrill and profane: "You lucky sonofabitch!" It took a moment to realize that the voice I had heard was my own. I was screaming at an American pilot, 2,500 miles away, who was a hundred feet above the earth and on his way out of the atmosphere.

Sheepishly, I looked around, half-expecting to find a curious crowd staring back at me. Or maybe a deputy sheriff, wondering why a guy would be parked in a car, alone, in the middle of nowhere, lips moving, yelling at demons. No one was in sight. I turned back to the radio.

A little before 7:00 A.M. on that morning of May 5, 1961, while I was setting out for work, Alan Shepard sat in a tiny Mercury capsule called Freedom 7 atop a Redstone rocket. He had just growled impatiently at his ground crew, "Why don't you guys fix your problems and light this candle?"

Driving along those quiet mountain roads, I had become conscious of how damn little I knew about the space program. It was very much in its infancy. The Russian Sputnik still circled above us, giving off those strange electronic beeps that resembled the sound track of an early Mickey Mouse movie. Sputnik had changed the direction of American science, had touched all our lives, and had brought Alan Shepard to that moment and that place—Launch Complex 5—at what was then (and would be again) Cape Canaveral.

I had read about the seven Mercury astronauts in *Life* magazine when they were selected, an American dream team. Instant celebrities. Lower-case gods. Not so much selected, I had thought, as blessed. At twenty-nine I was a marine fighter pilot in danger of becoming a career student at UCLA, and a physicist at the RAND Corporation, the famed think-tank. Nothing in my life had prepared me for the vicarious thrill of that incredible moment when Alan Shepard became America's first man in space.

It was one of those rare moments which, with just a tug of the memory, you can recall exactly where you were and what you were doing. The successful launching of Sputnik had caused a national trauma. Shepard's flight was to be the start of our comeback, and no one knew quite what to expect. Our unmanned rockets had been blowing up, keeling over, or cartwheeling into the ocean with appalling frequency. Millions hugged their radios and TV sets to share the exquisite suspense of that morning. Would Al Shepard be our first space hero or would they be picking up pieces of him over the entire length of the Eastern Seaboard?

At five minutes to lift-off the excitement had become too much, so I had stopped the car at the side of the road to just soak it all in. The damn announcer kept injecting his thoughts into the action when all I wanted were the lean, bare bones of what was happening. I wanted downs and yardage, not crowd color.

The announcer was embarrassed to be so excited. "Just think, an American is going into space on that rocket. I don't want to sound like a superpatriot, but doesn't that *almost* make you proud just to be an American?" Hell, yes!

At the instant of lift-off, I luxuriated in the feeling of the moment and then let out my announcement of envy. He had the most beautiful fifteen-minute ride ahead of him that any American had ever known. I wouldn't have recognized Alan Shepard if he had tapped on my car window, but I was thrilled by his adventure. I envied him as I had never envied any man.

My impression of Al Shepard has always been shaped by this first awareness of him. While his Mercury classmates each had their strong points, Al always seemed to me the most competent. He had won the contest to be the first American to ride a rocket into space. That's what the original competition was all about—to be the first to accept the challenge, to meet the test, to take the honors.

I recalled that spring morning on the Santa Monica mountaintop when, two years later, I competed for selection to the third generation of astronauts. The years in between had been filled with my immediate goals. To most Americans the space program was not quite real and the astronauts were a cross between Buck Rogers and James Bond. It had no more engaged my thoughts than the idea of volunteering to solve the political problems of the Middle East. But the envy was there with each lift-off.

One sunny weekday afternoon in June 1963, I confessed my feelings to Gerhard Schilling of the RAND Corporation. Gerhard had just returned from a year's loan to the National Aeronautics and Space Administration working for George Low, one of their top administrators. Gerhard wrote George of my relatively unique qualifications and a few days later a phone call from Washington got me started. A search was then underway for the third group of United States astronauts; the deadline for application was July 1, just two weeks away. In my résumé I emphasized my credentials for becoming a scientific member of the team. However, to my great joy, I found I also met all the qualifications for "regular" astronaut: United States citizen, less than thirty-four years of age, six feet or less in height, a degree in engineering or the physical sciences, and one thousand hours of jet pilot time.

Hell, if they had added hazel eyes and crewcut, the description couldn't have fit any snugger. There were nearly three hundred qualified applicants, including two women.

One other thing was needed: a recommendation from one's present employer. No sweat. My superiors at RAND were delighted to endorse one of their own for such celestial duty.

On the other hand my mother must have thought I was out of my mind. "What makes you think you can do something like that?" she asked.

I was a marine fighter pilot and within a year of finishing a thesis for a doctorate in physics; two tasks I had tackled because they were difficult and challenging. Becoming an astronaut was only reaching for one more star.

The summer of 1963 became a blur of forms, questionnaires, medical exams, screenings, briefings, technical interviews, and intermittent clinging to straws of gossip. From the hundreds who originally met the qualifications, thirty-two of us made it to the August semifinals: the physical examination at Brooks Air Force Base in San Antonio, Texas.

There we began to shape the curious, friends-and-rivals relationships that would be an essential part of the astronaut years.

The forty-five hours of activity at Brooks were roughly in three categories: health, physical fitness, and emotional. The breakdown is mine. It is possible to be in good health but in poor physical condition through lack of exercise. Mental state can be another matter entirely.

If a general observation can be made of thirty-two candidates from various backgrounds, I would say that we were, with rare exceptions, excellent in health, from good to excellent in physical condition, and well above average in intelligence. We shared a strong feeling of self-identity, confidence, and awareness of where we were going and how we would get there, and the competitive instinct surfaced quickly. There was very little interaction among the candidates, except in the evening—and then generally limited to one's three roommates. Whatever any of us lacked, it wasn't motivation. C. C. Williams stayed awake all night, bouncing up and down in order to insure being under the six-foot maximum height in the morning. He made it.

Some of the candidates weren't above trying to psych out the shrinks. One of my roommates checked out a textbook on Rorschach ink blot tests from the base library to determine the responses the psychiatrists were seeking (he didn't make it). While others engaged the good doctor in stare-down contests.

My appointment came several hours after having my eyes dilated, and I harassed the psychiatrist by wearing my sunglasses for the first hour of the meeting until he finally asked me to remove them. We sat there discussing subjects like *The Love and Fear of Flying*, a book written by another psychiatrist. It seemed my analyst endorsed the love-fear relationship in all aviators. I left just a little curious about how many, like myself, didn't fit his model.

All in all, the sessions with the shrink were interesting but a farce to most of us.

My self-evaluation was: excellent health, for which I really couldn't take any credit; good physical condition, which was surprising considering the sedentary college life I had led for the preceding eight years; and excellent mental and psychological form, which I had consciously worked at all my life.

But this was one time when the admonition "Know thyself" didn't take priority. You also had to know the competition. We were constantly measuring and comparing. They all looked good, and some (like Charlie Bassett) looked like shoo-ins. You would slip into the snack bar between tests to wolf down a bite of something—at lunchtime we were usually involved in a test—and find a Rusty Schweickart or a Mike Collins enjoying a diet of coffee and Boston cream pie.

What followed would not be a conversation in the usual sense; that is, an exchange of views. One might make a leading statement, tentatively, and hope that someone else would commit himself to an evaluation: "Man, that tilt table was interesting," or, "I hear some of the guys had trouble on the treadmill," or, "They say the shrink can ask some questions that are real beauts!"

One of my roommates at the Brooks Medical Center during the selection process was a handsome young navy lieutenant named Robert Schumacher. He was your typical All-American boy with a good flying background, and I conceded him better odds than myself of being selected. Surprisingly, he was one of four eliminated by the doctors at Brooks. Worse than that, it looked like a personal catastrophe since he departed the Medical Center grounded from flying. The doctors had detected, or suspected, a rare condition which violated the standards. Taking a conservative stance, they grounded Bob.

Schumacher returned to postgraduate school at Monterey, California, and spent the following year fighting to get his wings back. He was successful, was returned to a fighter squadron and in 1965 that squadron was deployed to the Far East on an aircraft carrier. Bob was the second U.S. pilot shot down over Vietnam, which I found out in 1966 sitting in my living room in Houston, Texas. There, on the evening news, was a North Vietnamese film showing captured U.S. airmen being paraded through the villages. At the front of the line, with his arms tied to a piece of bamboo behind his back, walked Bob Schumacher. It struck me then and a thousand more times in the next seven years how tenuous are the opportunities we have to significantly change our lives. While I was enjoying the fruits of the good life, Bob was rotting in Vietnamese prisons. He was also on my mind as we orbited high over Vietnam in October 1968.

Bob would not have been grounded had he never tried to become an astronaut, and he didn't make it only because of a complicated mistake. He returned home with the rest of the POWs in 1973 and is now back flying again. But who knows what might have been?

Two days after completing the week-long physical, I received orders with twenty-seven others to proceed to Houston for the final interviews. I had just passed Go!

The first week of September we went to Houston for further evaluation. By then I was more familiar with the qualifications and the backgrounds of the others and was consciously giving some candidates an edge. Dick Gordon had recently set a transcontinental speed record and was a retread on the selection process. Buzz Aldrin had shot down MIG's over Korea and written a Ph.D. thesis on space rendezvous. Half a dozen of them had been considered a year earlier for the second group—the Gemini Astronauts—had missed out, and were recycled to compete with our bunch. Another half dozen were big-time civilian test pilots.

It was easy to be impressed with the inside drag of a guy like Gordon. At a cocktail reception one evening Dick walked right up and spoke to Wally Schirra and Deke Slayton and Al Shepard. He *knew* them! *Personally!* My God, what chance did I have—those guys already knew each other! I not only didn't know anyone inside the program, but I didn't even know anyone on the outside trying to get in.

We were constantly calculating our chances of being cut. Wondering how much of the process we could control and how much was left to fate or blind luck.

Between the physical at Brooks and the final interviews in Houston, two of the candidates, Mike Adams and Dave Scott, narrowly missed death when their F-104 suffered a flameout during a landing approach. Each made diametrically opposed decisions and each got away with it. Mike, in the rear seat, ejected at the instant the plane crashed into the ground. Had he chosen not to eject, he would have been killed a few moments later when the engine of the F-104 slammed into the rear cockpit. Dave stayed with the plane and rode it out. He learned later that his ejection mechanism had been damaged in the crash, and he would have been killed had he fired the rocket cartridge. Each had considered such a situation in advance and carried out his pre-planned decision.

Dave Scott went on to the finals, was one of the fourteen selected, and had an eventful career as an astronaut. A back injury forced Adams out of the competition, but he was later selected as one of the first Manned Orbiting Laboratory (MOL) pilots for the air force. As the program dragged on and no one was getting into space, Mike resigned and went back to flying the X-15. He was killed in a high-altitude test of that winged rocket.

Nothing discouraged us. It was grand! It was glorious! It was patriotic! It was a challenge! It was man's greatest adventure and, by God, it meant a chance to be the first man on the moon.

Now there was nothing to do but wait—and worry. Slowly, not unlike college freshmen hoping to be pledged by the most snobbish fraternity, we formed our small alliances, sought out one another, pooled our insecurities, and—up to a point—lent each other support. I had a soul mate in Vance Brand. Vance was test-flying for Lockheed Aircraft Corporation in Palmdale, California, and seemed to have no better connections than I—that is, none. We roomed together in Houston and vowed to keep each other informed if any gossip came our way. I emerged from the three days in Houston feeling I was strongly in the running.

Flying back to Los Angeles, I found myself on the same flight with Dick Gordon and Ron Evans, who were returning to the Naval Post Graduate School in Monterey, California. They had a few hours layover in Los Angeles and, hungry to keep the talk going on what we had experienced or what might be ahead, I invited them out to the house to kill the spare time. Lo met us at the airport and we drove to the Valley.

Gordon did most of the talking. He is a good-looking, compact fellow with a breezy, James Cagney self-assurance, and he completely captivated Lo. There seemed to be no doubt in his mind that Dick Gordon, in a matter of weeks, would be a living, breathing spaceman. My own confidence began to wilt a little as I inhaled some of Gordon's exhaust. (Dick made it, of course, was twice in space and, in that unpredictable postorbital scramble, wound up as the general manager of the pro football New Orleans Saints.)

We tipped a few frozen daiquiris and talked for two hours of how our lives might be changed by admission to the space program. That week an item had appeared in the newspapers announcing the signing of a publishing contract between the astronauts and Time-Life and Field Enterprises. It was worth a half-million dollars annually to be shared equally among the astronauts (then sixteen in number, soon to be thirty). We were aware, for the first time, that more than just glory was at stake. From our standpoint it meant, at a minimum, financial comfort.

We agreed to share any early news on selection with each other, knowing full well that if NASA asked us to keep any news quiet, we damned sure would. At the airport we shook hands and I wished them luck, but only their fair share of it.

Through the friendly offices of the RAND Corporation I made one more trip to Houston. This time I was one of the company's two representatives at the summary conference for the Mercury program, only recently concluded, which had proved that man could survive in space.

I stayed at the Rice Hotel, one of Houston's landmarks. One afternoon I wandered into the Old Capitol Club with Wes Hooper, the sales director at the Rice. There we found Grissom, Shepard, and Paul Haney, who had succeeded Shorty Powers as the voice of the astronauts. They invited us to join them for a drink. In my search for positive signs, I took that purely social gesture as a favorable omen. I felt spacey.

But my first real burst of optimism came as the result of a phone call to Jack Cairl, in the personnel office at NASA. I wanted to know how soon we might get the word. "Soon," Jack answered. "Maybe a week. They'll notify those who haven't made it first."

Then, almost as an afterthought, Jack said that no decision had been made yet on what salaries would be offered any of the candidates who came in as civilians. "We're thinking in the thirteen-to-fifteen-thousand-dollar range," he said. "How does that sound to you? What's the least you'd come in for?"

What a question to ask someone who had never made more than $10,000 a year, and then only by pursuing three or four jobs simultaneously. I felt I would pay for the privilege, if I could afford it, and tried to sound casual. "Oh, thirteen-to-fifteen sounds OK to me," I mumbled. But even in that first inkling that I might be chosen for man's greatest adventure, I dared to think that I was being squeezed. "Hell," I thought, "I *know* being an astronaut is worth more than thirteen grand a year. Haven't they filtered the cream of American manhood? . . . for a forty-billion-dollar program?"

Returning to Santa Monica, I was still in the dark but more optimistic. That night I called Vance with my scuttlebutt.

A little over a week later, on the night of October 13, Vance called me. He was really down. "Walt? I just heard from Jack Cairl. . . . I didn't make it. What have you heard?"

I remembered what Cairl had said. Vance's disappointment, his sense of loss, came through clearly on the phone. I didn't know what to tell him and hoped I wouldn't have to share his hurt. My own odds were getting better.

The next day I sat in my office at the RAND Corporation, doing what I had done for the past ten days: waiting for the phone to ring and putting off work on my Ph.D. dissertation, the design and development of a triaxial search coil magnetometer which would measure fluctuations in the earth's magnetosphere.

And that day, October 14, 1963, the phone did ring. It was Deke Slayton, the director of flight crew operations. He was remembered as the Mercury astronaut grounded by a heart murmur, the only one who hadn't flown in space. It was an odd sensation to be able to say, with practiced familiarity, "Hello, Deke." (I've never been one to stand on ceremony, coming or going.)

Deke doesn't believe in dragging out the drama of a situation. He got right to the point. "Well, we'd like to have you come to Houston and join us, if you're still interested."

I hesitated for, oh, maybe a tenth of a second—I had hoped to complete work on my doctorate. "Yes, sir, when do I report?"

Deke told me to be in Houston on October 17. It was to be very hush-hush until the announcement press conference. I was to register at the Rice under the name of Max Peck. Years later I found out that all fourteen of us had been registered under the same name. Max Peck was the general manager of the hotel. I was to tell no one, other than my wife, that I had been selected. For a brief time, it looked like I wouldn't be telling her, either.

The phone at home was busy. And busy. And busy. I was bursting with the need to tell Lo, and I couldn't get through. It was frustrating. After more than an hour, she answered.

"Who have you been talking to all this time?" I snapped. (The biggest moment of my life, and I'm picking an argument.)

"I was taking a nap," she replied, sleepily, "and I took the phone off the hook. Why? What's the matter?"

"I made it, that's the matter," I shouted, my spirits up again. "Lo, I got the call from Deke Slayton. We're in. We're going to Houston."

Lo was awake now. She caught my excitement. Soon we were giggling over the phone like school kids.

"I'm not supposed to say anything," I said, lowering my voice, "but if anyone walks in and sees the look on my face, they'll sure as hell know."

We had two days to savor the news, to think about the implications and mentally rearrange our lives. A new career. New city. New friends. A new environment for our kids: Brian, then two, and Kimberly, still in diapers. We both recognized going in that it was a fantastic opportunity. Not just being an astronaut, but for what it offered in the way of a future afterwards. I didn't apply for the space program to make it a career, but with the expectation of devoting a reasonable period of my life to it and then stepping out and going on to other things.

Flying alone to Houston on October 17 I gave myself a pep talk all the way down. I thought about my credentials: five and a half years of active military service, eight more of schooling and research. Twelve years a fighter pilot, I considered myself a skilled aviator—the best. But I'd be competing now with thirteen pros, half of them test pilots. "This is the big league," I told myself. "You can't let up. You've got to keep charging." For a goal-oriented man it was one more mountain to climb.

The Apollo astronauts were introduced to the world and each other on October 18, 1963, exactly five years before my first and only space flight. We were an interesting mix: seven from the air force (Buzz Aldrin, Bill Anders, Donn Eisele, Charlie Bassett, Ted Freeman, Dave Scott, and Mike Collins); four from the navy (Dick Gordon, Gene Cernan, Alan Bean, and Roger Chaffee); one marine (C. C. Williams); and two civilians (Rusty Schweickart, a research scientist from MIT, and myself). Rusty had just resigned from the Massachusetts Air National Guard, but I retained my Marine Air Reserve affiliation.

Our group was to represent a cross-section of the Apollo program itself, with all the highs and lows of astronaut careers. Aldrin and Collins would fly the first lunar landing mission. Chaffee would die in the fire on the pad. Bassett, Freeman, and Williams would be killed in jet crashes. Eisele would become the first divorced astro. Scott would be at the center of a scandal for taking envelopes to the moon.

We were entertained at a cocktail party the night we arrived, and some of the older hands dropped by to welcome us. I learned later that they did not attend such affairs happily. They were busy traveling, working on the Gemini program, and it was an imposition to intrude on one of their rare free nights. Still, we visited with Deke Slayton and Al Shepard. Wally Schirra and John Glenn, the reigning hero, showed up along with Pete Conrad, Jim Lovell, Jim McDivitt, and a few others from the second group. Their poise and sharpness were dazzling. Under the circumstances it was easy to overrate them. I wasn't familiar enough yet to know their weaknesses. They were all impressive. The Gemini astronauts seemed better equipped and more rounded, though they lacked the charisma of the Original Seven. They hadn't tasted the glory of space flight and were still unknown quantities.

Dr. Robert Voas, a psychologist who was the Mercury astronauts' training director, detailed the required characteristics as he saw them: intelligence without genius, knowledge without inflexibility, a high degree of skill without overtraining, fear but not cowardice, bravery without foolhardiness, self-confidence without egotism, physical fitness without being muscle-bound, a preference for participatory over spectator sports, frankness without blabbermouthing, enjoyment of life without excess, humor without disproportion, fast reflexes without panic in a crisis.

Against that backdrop we felt a little like a rookie football player out of, say, Bethune-Cookman College at his first training camp with the Green Bay Packers. We were grateful just to be there and in some ways awed by it. We flat bathed ourselves in the feeling of being on the team. We were a little nervous, and, at the same time, cocky.

We were in, and there was just one public hurdle to face. Our first press conference. Paul Haney briefed us before we met the newsmen and fed us sample questions. We were assured that the questions would be simple and friendly: "Were you a Boy Scout? What is your religion?" Paul added, "The only problem might come from questions about the personal-stories publishing contract. You can honestly answer you don't know much about it yet."

After the briefing, Rusty Schweickart and I hung back. We cornered Paul and, as it quickly developed, we both shared some concern for the same potential question. "Paul," Rusty blurted out, "I don't know how you'll react to this, but I'm not what you'd call a religious person. I can't claim any religion."

It was my turn to chime in and add to Paul's problems. I had abandoned organized religion fifteen years before, and was at best an agnostic in most people's eyes. Rusty and I weren't amused by our mutual dilemma. We weren't taking it lightly. The American public (and NASA) might not look with favor on sending astros without a firm belief in God on a mission deep into the heavens—that close to the home office, so to speak.

And it was no frivolous point. When Yuri Gagarin returned from his first orbital flight, he made a point of sniping at the spiritual attitude that has been so much a part of the American ethic. "I didn't see God up there," he cracked. "I looked and looked and looked and I didn't see God." The American public was shocked at and repelled by Gagarin's blasphemy, and cited it as one more example of "godless Communism."

We only had a few moments, but Haney betrayed no real concern and responded matter-of-factly. "If they ask, just hedge. Tell them your family's faith, or something." Although a devout Catholic, Paul didn't presume to judge us, nor did the disclosure seem to disturb him.

The reporters' questions were trivial enough. Then, addressing all of us, one of the reporters asked, "What is your religious preference?"

We responded one by one, starting at the tail end of our alphabetical seating. When they reached Rusty he replied calmly, "I have no preference," and the questioning continued down the line. No one thought anything about it.

When my turn came I copped out. I couldn't bite the bullet, and Rusty's answer was too perfectly conceived to copy. Being candid isn't always a virtue, and I said, "My mother and sisters are all Lutheran, but I've gone to many different churches."

The press conference ended without incident. The next day's Houston papers ran our photographs with long strings of words under them. It was official. We were the third generation of astronauts. We were the men of Apollo.

I thought I knew what NASA had to offer me. But even with all the security checking—questioning neighbors, friends, classmates, coworkers—I wondered if they really knew what they were getting in Ronnie Walter Cunningham.

Mine wasn't a remarkable story. Up from poor; oldest of five kids; an egghead in school who struggled to be recognized for his athletic endeavors; I went to work at eleven delivering papers and have paid my own way ever since.

My ticket to upward mobility was ambition; the vehicle was the airplane. At ten, brainwashed by the movie *The Hell Divers*, I resolved on the spot to become a lieutenant commander in the Navy Air Corps. Between the fourth and sixth grades, my claim to fame was being able to identify aircraft flying overhead by the sound of their motors. At the first distant hum of an engine, the teacher would look at me. I'd cock an ear and announce, "P-thirty-eight out of Lockheed," or "A-twenty just taking off from the Douglas plant." In 1941 and 1942 southern California had a lot of aircraft in the air.

It was the era of the Quiz Kids, and I competed unsuccessfully in local eliminations to qualify for that precocious panel. Still, the reputation of being a brain haunted me all through school. From accelerated classes in grammar school through high school, I carried what seemed the double burdens of being both an intellect and a runt. My growth didn't come till after high school.

I was athletic and did well in gymnastics. In one of the team pictures I was the only guy standing and was still the shortest one in the picture, five foot four and wiry—or "lean and mean," as I characterized it in later years.

I was a smart punk who made A's and generally had his nose in Mickey Spillane's latest thriller. I could knock off one of his paperbacks in two days just reading as I walked between classes.

My school years were a constant struggle to become one of the guys, to hang around with the school toughs, and in general to make a conspicuous effort to be something I wasn't.

Mother was ambitious for her children and encouraged my studies. Dad had a small cement contracting business, but he liked being on his own primarily for the freedom it gave him. He dearly loved to fish and hunt with his cronies. That was his declaration of independence.

From out of that I somehow developed the drive to set high goals and achieve them. I grew up with the notion that with hard work I could do better than the next guy.

I enlisted in the navy in 1951, without the two years of college required for pilot training. But a year and a half later I passed an equivalency test and was accepted for Preflight School at Pensacola, Florida. I gave myself the nose-to-the-grindstone speech, and it took for about six weeks. The posting of the first "grades" showed me second in our class of fifty-six, and characteristically, I coasted for the rest of Preflight.

When we moved into the flying phase of our instruction, my tendency to get by on less than 100 percent effort was shed immediately. Nothing had ever come to me more easily or naturally than flying. It was as if Walt Cunningham had been reborn. I felt like van Gogh picking up his first paintbrush. From the beginning, I lived, grew, belonged, in an airplane. I loved it, and for reasons I only began to understand years later.

The mechanical skills are only a small part of what makes an aviator. It's what's in the head and heart that makes a great pilot. Flying single-seat, high-performance airplanes, is where it's really at for those who *know* they belong above the clouds. In those days I viewed flying as a joyous, liberating experience. I was ecstatic when I whipped my instructor in mock combat—and it seemed a natural thing to do.

What I didn't realize was how this phase of my life was shaping me for the future. In time, I came to know that the magic for me was in the command and mastery one has over one's destiny when flying an airplane. It hit me first on a night formation flight, in advanced training, in an F6F Hellcat. What developed was a rock-solid case of vertigo. While flying above the Corpus Christi area in Texas, my senses lied to me, they said we'd been flying formation in a near-ninety-degree bank for the last ten minutes. The instructor flying in the lead blinked his lights, signaling me to slide over between his aircraft and the Number 3 man to form an echelon, with all four aircraft in a single line. In the second it took to move across, I thought, "My God, he trusts me—giving me a crossover in a ninety-degree bank. The slightest error on my part and we'll both be dead!"

There was *no* way I'd make that error, and zap, I was in position. It wasn't a good crossover—too fast—but it shook me out of the vertigo, and there we were, as we had been all along, flying straight and level. That night I came to the realization that I was carrying my life in my own hands and it has stayed with me ever since. In fact, the entire pattern of my life was set by that pilot training. It has caused me to confidently seek a challenge wherever I can find one, to charge ahead and never look back. That opportunity to achieve a feeling of mastery and control over oneself, that feeling of omnipotence, is worth all the effort it takes to get there.

In another sense, flying was to me a maturing, exciting way to move from my rather modest childhood toward the good life. It was a filtering process that marks you a little unique. In retrospect, it was also the first screening toward becoming an astronaut and flying in space.

After serving as a Marine Corps pilot, and getting married, I enrolled in college. We lived on the GI Bill and Lo's salary as a secretary. Concentrating on school, and holding no job that first year, I was strongly motivated to produce results to justify Lo's efforts to support us. If anything, I overcompensated, finishing the first several years with straight A's. After taking a job and adjusting priorities, the old characteristic took over: laziness. I worked hard on mathematics and physics and settled for average grades in everything else.

That strategy had carried me through my military and college careers. Now I looked on the space program as the ultimate testing ground. Whatever my talents, whatever my capacities, they would surely be tested and found out here.

We plebes had reporting dates for around January 1, 1964. Rusty Schweickart showed up before Christmas, which made him an old hand when I arrived on January 3. The two of us shared an office in a reconditioned building—formerly condemned—at Ellington Air Force Base, eighteen miles southeast of Houston.

Our indoctrination, education, and some disillusionment were still ahead of us. Soon after the group assembled, we were treated to a memorable lecture by Wally Schirra, introducing such subjects as Astronaut Deportment, Fringe Benefits, and How to Handle Per Diem Checks and Travel Expenses. Wally offered such pearls of wisdom as suggesting each of us open a travel account into which our per diem checks would be deposited. This would allow us to keep track of expenses and simplify the handling of business deductions at income tax time. It was also the first indication we had that being an astro would involve interminable travel and that our finances were apt to get complicated, and—for some—best kept private.

Wally's lecture also contained smiling references to something that was already becoming obvious to us. The ladies just loved the astronauts. It was wild, the lengths they would go to to be . . . friendly. We were expected to tolerate a certain amount of freewheeling behavior on the part of our compadres.

With the astros the biggest problem was loose women. The opportunities and the temptations were fantastic. C. C. Williams, the first bachelor astronaut, had someone from Lubbock, Texas, chasing him with a paternity suit by the time he reported to NASA. It wasn't too surprising, in light of later stories that never came off, to learn that C. C. had never been in Lubbock. Some dude had passed through town saying he was C. C. Williams, famous birdman, and left a girl pregnant.

The sense of what Wally told us was to cool it around the flagpole. There were going to be goodies we never dreamed of—good deals on Corvettes, 4 percent home loans—and we, the weenie astronauts, would be cut in by the old hands. In return we were expected not to rock the boat, not to make waves by repeating indiscreet remarks to the press, neighbors, or our wives.

Of the thirty astros then in the program, twenty-six were career military men. (The four civilians were Neil Armstrong, Elliott See, Rusty Schweickart, and myself.) And even though NASA was a civilian agency, the Astronaut Office had a definite military cast. I felt as though I had been recalled to active duty with a fighter squadron. The

Astronaut Office, which was headed during most of the sixties by Al Shepard, operated on an obvious "Rank Has Its Privilege" philosophy.

Yet, as in most things involving the astros, a paradox existed. Few wore uniforms even once a year (as a reservist, flying weekends with the Marine Corps, I was in uniform a helluva lot more than the career men). Today, most would feel out of place and at some disadvantage if called back to their parent service. Col. Buzz Aldrin returned voluntarily to the air force and lasted only a year. Gen. Tom Stafford has done much better.

More important than service rank was the pecking order: whether you were first-, second- or third-generation astro. We were the new guys, the rookies, low men on the totem pole.

I had reported with an open mind toward the experience but a little sensitive about my flying background—not having been a test pilot, although half of those selected were in the same boat. I was a civilian marine reserve officer at whom regular career officers tended to look down their noses. Coming from an academic campus background did little to change the picture. So I bent over backwards to show I was a tiger, aggressive, competitive and the best qualified of the bunch—not without creating some problems.

From the beginning Rusty and I suspected we had been selected as sops to the scientific community, but it took that august group, the National Academy of Sciences, only a little while to find out that we were just like the rest of the "dumb fighter jocks." Flying was much more attractive than working out a problem in physics.

The Astronaut Office was one of three divisions—the others were Flight Operations and Flight Crew Support—in the directorate headed by Deke Slayton. His appointment as deputy director had been (at least partially) a consolation prize bestowed on Deke by the other Mercury astros when the medics had grounded him. Expecting that Deke would never make a space flight, the Mercury astros relinquished a certain amount of their autonomy by "allowing" him to be in charge. He lacked management experience, but started right off running the show. He was successful because he had their respect. He was one of them; he had gone through the same selection process, the same trials. This was no outsider trying to issue orders.

Many times his authority required a delicate hand. It was never exactly a jelly roll to be "in charge of" Gordon Cooper, Scott Carpenter, Alan Shepard, Wally Schirra, or Gus Grissom. John Glenn, of course, was soon to be leaving the program. In some respects, the boys gave the impression that it was "Deke Slayton, Director of Flight Crew Operations, by the authority of the Manned Spacecraft Center and the grace of the Mercury astronauts."

Within weeks of our arrival we were on a full schedule of orientation, classes, and training. It was a demanding, self-consuming job. Waking up in the morning, my first thought was always on the day's activities.

The general training was also worthwhile as an exposure to each other's strengths and weaknesses. Flying around the country with different astros provided an opportunity to learn who felt there was only one way to do things, their way, and who was more flexible and tolerant of different points of view.

The routine, the training, in a sense, our careers were loaded with ironies. After several flights with Ted Freeman, I was convinced he was one of the better pilots I had ever known. Only a few weeks later, Ted became the first astronaut to die . . . in a jet crash.

The warmest and most sincere receptions from any of the Mercury astros were provided by Scott Carpenter and Gordon Cooper. They were the only ones eventually frozen out by their contemporaries.

From the outset there were signs of the frustration to come. We had no clue as to when our flights would be projected or who would be selected to make them. In a corner of my mind was the nagging thought that maybe all this was simply to keep us busy while the Mercury and Gemini guys got ready and flew their missions.

Our uncertainty was aggravated by the fact that we were attending "gentlemen's classes"—no grading. I had looked forward to, had actually craved, the intense competition that the public suspected was encountered on every level. If I was going up against twenty-nine other aces, I wanted it measured, taped, recorded—and wanted to see it affect each man's progress. The reason for "gentlemen's classes" was simply that the old heads wanted it that way. The Mercury astros refused to be graded on anything they ever did. They were reluctant to leave a record showing that Wally Schirra was better (or worse) than Scott Carpenter or Gordon Cooper. They sat at the top of the hierarchy and could only lose by laying it on the line to be measured.

In the end it didn't make a nickel's worth of difference how we performed in class. Take geology, which not all of the fellows could. We had lectures, at first two or three times a week, later two or three afternoons a month. You showed up if you wanted to, and the new guys always did—at least until we got a bit salty or grew tired of looking at rocks. Out of the Mercury group, only Carpenter and Cooper made a consistent effort to attend. Deke Slayton was up to his armpits in paperwork, but he made a point of showing up when he could. Deke was trying to set an example, and he kept hoping that he would be returned to flight status.

We weenie astros attended faithfully for one good reason: we were looking desperately for an edge. In terms of making an impression, we had precious few options. We didn't get to fly around the country with

Deke, selling ourselves. We were spending most of our time in classes
—staring at a blackboard or attempting to sort out a box of rocks. The
professors had zero influence, and there weren't many dramatic possibilities in that situation: no safely bringing down a burning aircraft, no
untying little Nell from the railroad tracks.

It quickly became apparent that in order to make out in the program
—to get in line for a flight crew—you needed the right game plan. Bill
Anders and I shared one based on the hard-work ethic: bust your ass,
do your best, don't let up, and the boss will eventually notice and
reward you accordingly. We stuck to that for three or four years. It
wasn't the best plan.

Everywhere we went, this was the major topic of conversation. In
late 1965, most of us traveled up to Boston for an introductory course
at Massachusetts Institute of Technology on the Apollo Computer and
Inertial Guidance System. One night Anders, Schweickart, Bean, and
I—we usually palled around together—went to the famed Concord
Bridge, where the Minutemen fired "the shot heard round the world."
But instead of being inspired by the history of the place, instead of
talking about the Revolution, we stood there talking about space and
our chances to get into it. Another night we drove over to Harvard to
poke around. (I was curious to see the famed Harvard Yard.) And
again, while sitting in a small café on campus, we talked about how to
get on a flight crew.

Some of our guys (Mike Collins, Dave Scott, and Dick Gordon)
were already assigned to Gemini backup crews, and we quickly began
to discuss the mysteries of crew selection. By what process did the
magic wand reach out and tap you? Who would get the first Apollo
flight? Who would fly to the moon? What could one do to enhance his
chances of being chosen?

"It doesn't work; it's not enough," Anders finally decided.

"What isn't?" I asked.

"Hard work and doing a good job. No one really notices."

We were constantly analyzing the system and playing the name-the-
crew game, over coffee or a drink or when we were rooming together
on the road and the lights were out and we couldn't sleep.

We were slowly waking up to the fact that politics and favoritism
were very important. It wasn't that much different from any other job
where personalities play a big part, where it helps to be in the right
place at the right time, and where certain factors—service relationships, first impressions, and pressure from friends (pro and con)—
created fair-haired boys.

It wasn't necessarily a power play. It might work like this: Deke
Slayton had worked with Frank Borman in the air force, and Borman
had worked with Mike Collins. When it came time to move the first

Apollo astronauts into flight positions, Mike Collins would be selected to back up Gemini 7—which happened to be Frank Borman's mission. Later, Collins would move on to Borman's Apollo 8 crew.

Another example: Dick Gordon (who had been friends with Pete Conrad since navy test-pilot days) was paired with Pete as backup on Gemini 8. From there the two of them flew Gemini 11, backed up Apollo 9, flew Apollo 12, and went on to back up Apollo 15. Al Bean, who also knew Pete and Dick from Naval Test Pilot School, joined them as the third member of their crew after the Gemini program.

It paid to have a champion in the hierarchy above you. Once in the rotation, the chances were strong of being recycled on still other missions, leading, one hoped, to that pyramid in the sky: the lunar landing.

The days flew by and the frustrations mounted. There was seldom, if ever, any feedback on how we were faring: good, bad, or indifferent. We slowly got fed into the paper mill, with directives and memoranda and meeting notices constantly coming across our desks. We worked our butts off on the Gemini program even if we were quite certain we'd never make a Gemini flight. We never knew who was watching and making what kind of decision based on what information. It was an open-loop competition at all times—no feedback to the contestants.

We went to class and scrambled to get a few minutes in the Mercury simulator. We hustled to get an early checkout in helicopters. In our ignorance or naïveté, everything or anything could have had some importance, even the dullest classroom lectures. It wasn't just that someday our lives might depend on some piece of casual information, we paid attention because someday, somewhere, someone in a meeting might ask a question. If we could summon forth that one jewel of an answer, retained from hours of drudgery, then bingo, our lives would move out of the shadows and into the center arena.

All the time we were competing with masters in the art of gamesmanship. One never admitted that a drill had been difficult, that he was encountering problems, or that he *ever* felt insecure. We were always doing "great, just great."

One of our early brushes with "playing" astronaut was the Mercury simulator at Ellington. (The Gemini model barely arrived in time for the first Gemini mission and simply wasn't available for practice licks by guys who weren't even in line.) The challenge in the Mercury simulator was to fire the retro rockets to bring the spacecraft out of orbit with bad misalignments in all three rockets. (There was a serious aftermath to this exercise. Scott Carpenter landed his Mercury capsule 350 miles down-range as a result of not adequately controlling the vehicle during retro fire.)

Three rockets were fired, one after the other, and the object was to

hold the spacecraft precisely on attitude during the burn. Unless somebody came and looked over our shoulders, which we resented, or unless the clerk at the training table told someone, no one would know
how well or how poorly we did. I might discuss the task in general
terms with Charlie Bassett, but if he got curious as to how I was doing,
"It was a piece of cake."

3

Taking the Bad With the Good

I DON'T BELIEVE it would be an overstatement to say that we were in endless conflict with our own egos. No matter how hard we might try not to take ourselves seriously—and some tried harder than others—we were constantly reminded of our special role as men being trained to leave this planet.

There were many models for us to choose from: John Glenn, with his sense of obligation and higher purpose. Tough, intense, cocky Gus Grissom, the kind of guy who didn't care if the sun came out or not. Scott Carpenter and Gordon Cooper, both adventuresome, who were willing to pay the price—and did.

And Al Shepard. I suspect that even the most cynical of us admired the sophistication, the social poise he brought to the job. Al was everybody's friend. Governors knew him. Millionaire oilmen knew him. Kings-in-exile knew him. His technique was flawless.

Our own situation—so near and yet so far—was apparent to us even on such exercises as our jungle training expedition in Panama in July 1964. We hacked our way through the jungles for three days, two groups of us, enduring the humid heat and underbrush and mosquitoes that could carry off a small dog. After a half day's walking, we were paddling our way down the river in two dugout canoes when Pete Conrad glanced up and happened to spot a lean-to hut squatting on a hilltop.

"Hey, how about that," yelped Pete, pointing. "Way out here in the middle of nowhere, in the middle of the jungle, a little old hut up there with a family living in it."

"Yeah," said Jim Lovell, drily, "and hanging on a wall in there someplace is a photograph signed, 'Warmest Personal Regards, Al.'"

That broke us up, even as we glided along the backwaters of the Panama jungle. Shepard, of course, was famous for his photos with that stock phrase. I have one on my own office wall with the same inscription. When you have to sign them by the thousands, a fellow can't be blamed for using little imagination.

In an entirely different spirit, another astronaut trait became evident early in our training: knocking the other guy. In eight years at NASA, I rarely heard one astro volunteer anything complimentary about another or admit to his own problems. The one consistent exception was a postflight acknowledgment of a mission professionally conducted—if it was truly warranted.

It was disturbing to know that your friends and co-workers thought nothing of putting you down in the eyes of the NASA brass, most notably Al Shepard and Deke Slayton, who could tap your shoulder and say, "You're gonna make a space flight."

The putdowns might deal with behavior or office style ("Cunningham? He can't get much done with that phone stuck in his ear"). Or might plant a seed of doubt about a buddy's flying competence ("Have you noticed Al Bean rarely flies at night anymore?" "That Bruce McCandless never flies in formation?").

Under the circumstances it wasn't always easy to remain friendly. The more open you were, the more exposed you became to snipers.

It wasn't so much out-and-out cutthroat as it was a spirit of laissez-faire, free enterprise, anything goes, win-at-any-cost competition—all accepted values in the fields of sport, commerce, and politics. We felt more confused and isolated than bitter. There was no handle to be found, no real understanding of how the machine operated and how to deal with it.

With contacts in the first or second groups one might pick up a fragment of useful information now and then. But I hadn't known any of the other twenty-nine astronauts before joining NASA and none of them had any reason to know me. So my intelligence loop revolved around the likes of Anders and Bean and Schweickart. We were terrific at gossiping and forming opinions, but we had few facts to exchange. We were, even collectively, a fairly useless intelligence network.

There was one respect in which the program was eminently fair. Undesirable assignments were distributed, without prejudice, to all.

A notable example of what was considered undesirable was what we called "the week in the barrel." The inspiration for the name came from the old story about the sourdough who staggers into an Alaskan mining camp, looking for a woman—any woman. He has rather serious biological needs. And they refer him to a back room that features a large barrel with a hole in the side.

The week in the barrel was the result of an arrangement between the Astronaut Office in Houston and NASA headquarters in Washington, D.C. Every other week our office would deliver one astronaut body to NASA headquarters to brief, schedule, and use as they wished. It was a public relations ploy designed to help NASA sell the program to the taxpayers and congressmen of the United States. The agency had literally thousands of requests a year for us to speak to everything from kindergarten classes to Congress (Congress got priority). On a rotating basis, starting with the *A*'s (for Armstrong), each astronaut was expected to spend one week on what amounted to a whirlwind speaking tour.

For those of us who had yet to get a taste of space—who were, in fact, still years away from going up—the week in the barrel was an embarrassment. I never got used to the idea of a roomful of citizens giving me a standing ovation, when all I had done was fly in from Houston. The most exciting thing to talk about firsthand was the trip to Panama and the week of jungle survival training. We looked forward eagerly to the arrival of a newly selected group of astros, simply because it meant more time between weeks in the barrel.

My first such week began with a briefing in Washington, where James Webb, the NASA administrator, Julian Scheer, then the director of public affairs, and other big guns helped me bone up on the program. Then, starting with an audience of four hundred engineers in Schenectady, New York, I went on to make thirty appearances—or, more correctly, the same speech thirty times—in five days.

After Schenectady I made a swing through Alaska, hitting every high school and service club in Anchorage and Fairbanks. The week was an education, starting with the VIP treatment; NASA assigned a protocol officer—in those days the agency could still afford it—to keep me on schedule and out of trouble. Mine was a cool twenty-four-year-old named Fred Asselin who had a splendid instinct for knowing when and how to get you out of a crowd and away from autograph seekers without making you the bad guy. He had moxie and knew how to play the game. The mayor of Philadelphia listened meekly as Fred assigned automobile priorities to the dignitaries when the city honored Al Shepard with a parade.

These speaking tours provided most of us with our first real glimpse of the adulation lavished upon anyone who wore the astronaut label. It was late November 1964 when we landed in Anchorage and it was 0° Fahrenheit outside. The coldest weather I had ever experienced previously was 12° *above* zero during a bleak Korean winter, and since I am from southern California, I freeze my tassels when the thermometer drops below 70°.

After a full day of listening to myself orate—I had a ten-minute film and a stock spiel about the training grind, the only thing I knew about—we had a midnight flight to Fairbanks. When we climbed out of the plane, it was *thirty below zero.* I thought I was going to faint. And there's a crowd on the runway to meet me, as though I had just flown the Atlantic in a bathtub or something.

Someone lent me a fancy wolverine parka, and it sure came in handy the next day. The local school kids were excused from class in the morning for my appearance at the air force base. I stood in an open Jeep and waved at the kids in a kind of three-block miniparade. It was so cold! Even today, I shiver when I think about it.

But the incident that made my week in the barrel was a classic case of astronaut overextension. That's a combination of what other people think we can do, and our willingness to try it anyway.

To keep our appointments, the commanding general of the Alaska Frontier had made available his personal helicopter: an HU-1A, or Huey, made famous in Vietnam rescue and support missions. It was a turbine chopper with a deluxe interior, all dolled up for the big brass. The pilot was a bright, friendly, obliging army captain by the name of Toso.

Having bootlegged five or six hours of stick time in helicopters, I was looking forward to checking out in them. Over the noisy clatter of the chopper I yelled at him, "I'm going to helicopter training in February," and again, louder, "I'm going to helicopter training."

He smiled and nodded. After a bit he asked if I'd like to take the controls. I accepted with alacrity. After all, I was an astronaut; I could do anything. A moment later I was flying the general's helicopter. I was lifting off, hovering, flying to the next speech, and we're both having a damn good time.

Later, we had ten minutes between speeches, and the captain offered to let me practice on their training course nestled between the trees on an old airport. I leapt at the chance. We had Fred and another passenger in the back seat and I wasn't doing badly for a guy with maybe six hours of flight time in choppers.

The ground was blanketed with snow and I was threading my way around the course, with trees on either side of me. My confidence was soaring. Then I looked up and saw a telephone line directly in front of me. With that fabled reaction of the trained aviator, I realized there were two choices: I could go up and hit the wire with the canopy or go down and cut it with the rotor blade.

With unfailing judgment, I took a third choice, turned to the captain, and said, "It's all yours." He went right through it with the rotor, cutting the lines and wrapping them neatly around the mast. We were only about five feet off the ground and he sat the chopper down smoothly.

I sat there, feeling stupid. He looked at me, hesitated, and asked, "When did you say you went through chopper training?"

"Next February," I said.

"My God," he said. "I thought you had already finished it."

We tried and were unable to untangle the telephone wire from the rotor blades. It was 4:30 P.M., getting dark and growing colder. We were unable to raise anyone on the radio and it was looking grim. Finally, an army observation plane spotted us and called in a spare helicopter. I ended up missing one speech and getting a lesson in the folly of believing your own press clippings.

I really felt bad about it—not only because of my own embarrassment, but because my friend, the captain, was responsible for the general's personal helicopter. I wrote letters to the general and to everyone else connected with the trip and, luckily, there were no repercussions. I learned another lesson: no one was anxious to take potshots at a working astronaut.

Most of the time we were working our tails off. We existed for those moments when, whatever the reason, we were singled out from the pack. My first such opportunity came when I was assigned to test a new spacesuit on a field trip to the McKenzie Lava Flows in Oregon. It was no big deal. I may have been chosen because the only suit available was in my size or because everyone else was busy. But at least it was something I wouldn't be doing in a group.

The test turned out to be difficult, tedious, and tiring. I struggled and scrambled to climb a mountain the likes of which no one would ever encounter, much less climb, on the moon. In the process one day I slipped and fell against a sharp rock, punching a tiny hole in one glove of my pressurized suit. It caused a small leak and the suit wouldn't hold pressure, so we had to send for a spare glove.

While all this was going on, photographers with telescopic lenses were zeroing in from fifty yards away. The next day all the papers were full of Cunningham falling on the lava bed and ripping a glove.

The test was successful, and we accomplished what we had set out to with the suit. It had been hard work, but I was satisfied. On returning to Houston the only thing Deke said to me was, "Walt, you shouldn't have fallen down like that in front of the news media."

The trip did result in one nice surprise, though. While I was out there on the lava beds, a string of cars pulled up. Out of one of them stepped Mark Hatfield, then the governor of Oregon. He introduced himself and we shook hands, posed for pictures, and quickly finished the political amenities. While we were making small talk, he mentioned that the local pilot who had flown me in from Klamath Falls had reported I was going to put in the new house I was building a lava rock wall to represent the lunar landscape. The upshot was that a few days later the governor had some lava rock crated up and shipped to me, enough to finish the end wall in my living room. It made a nice conversation piece for years.

During this early training we welcomed any kind of individual assignment, as a capsule communicator (CapCom) at Mission Control—at least you got to talk to the *real* astronauts—or in a mission support role at the Cape.

It was a period in our lives when, every time we went on a trip and shared a room with a close friend, we'd lie in bed at night and let out whatever was inside us. Unless one were on a flight crew, he was predictably mad as hell about the system, and we bitched about crew selections and complained of the trapped feeling. We were the victims of the judgment of some nebulous higher authority, and it was frustrating. No one ever expressed concern about failing, only about not getting an opportunity.

It was like the famous line by Henry Jordan, the Green Bay tackle, about Vince Lombardi: "Coach is fair. He treats us all alike. He treats us all like dogs." The system was fair in that respect. Nobody told you anything, nobody promised you a damn thing.

We all went through those emotional bends at one point or another. I drove home from Ellington Air Force Base one day with Pete Conrad, then training for his Gemini 5 "eight-days-or-bust" flight with Gordon Cooper. Pete was tooling his Corvette down the Galveston highway, batting along at 90-plus and blowing off steam. He had decided that the space program wasn't all that neat, after all. He was going to hang up his jock. He had it all figured out.

"You can't quit this damn business," he said, dead serious, "until after your second flight. If you quit after the first one, they'll say it scared you to death, and you didn't want to go back. So you have to make a second one. But after that, they can have it. There are other things a guy can do besides taking all this bullshit."

I was sitting there agreeing with him—but on an entirely different level.

The kicker, of course, is that in June 1973, eight years later, Pete Conrad commanded the first Skylab mission, his fourth space flight. When it came to leaving NASA, there was a wide disparity between promise and performance.

At least 75 percent of our time during the first eighteen to twenty-four months was spent on an organized training schedule. When we weren't in classes, on field trips, or in the simulators, we were flying, often in pairs with other astros. It was natural to begin calibrating them. My first ride with Elliott See destroyed any visions I might have harbored about the super-engineer-test-pilot types.

Elliott had flown the T-38 as a test pilot for General Electric long before he came to NASA. On this particular flight he was up front, with me in the rear. Immediately after takeoff—while we were still heavy with fuel—he indicated he wanted to fly over Taylor Lake and buzz his house, a not uncommon astronaut practice. Elliott had told his wife, Marilyn, when to expect him.

I didn't object. A buzz job to me was fun; it meant coming down faster than a bat out of hell and saying "adios" before anyone gets your tail number. I have flown landing gear checks faster than this particular low pass. Elliott dropped down slow and got set up to cross Taylor Lake at fifty feet with a full load of fuel, at 170 knots, and no flaps down. For the uninitiated, let me inject here that the T-38 with a full load of fuel won't fly a whole lot slower than 165 knots, even with the flaps down. I was in the rear cockpit, rationalizing that he was a good aviator, he had to know what he was doing. After all, he was one of the big boys. The fact that he was flying way too slow troubled me only to the extent that it could get me killed.

In ninety-nine out of a hundred situations, the second pilot does nothing but try to anticipate where the boundary is when you both must suffer the consequences. Deciding we had passed that boundary, I finally piped up and said, as casually as I could, "Hey, how about half flaps?" He dropped half flaps, waved to Marilyn, and we flew away. Unfortunately, that episode flashed to mind when Elliott and his seat-mate, Charlie Bassett, were killed the following year in a low, slow landing approach at St. Louis that ended in a stall and crash.

Elliott was a great guy, a friend of mine. But when judging another pilot's capability, someone on whom your own life might someday depend, you can't afford to get sentimental.

In late 1964, as we neared the end of our formal training, Deke dropped an overnight assignment on us called a "peer rating." It was a procedure that had been followed by both the Mercury and Gemini groups. The peer rating process was familiar to the fellows who had been through Annapolis and West Point, but to the rest of us it inspired only a blank stare.

Said Deke, patiently, "You simply eliminate yourself and then consider in what order you think the other guys in your group ought to be flying a space mission."

Deke advised us to think about it overnight. In effect, we were being asked to speculate about each other's performances. I went home that evening and thought about it. I suspected there were various bits of gamesmanship that could be played with such a list. Some might have been tempted to put the softest competition at the top on the chance that Deke would allow so many points per vote, like football ratings. But even that theory required a double-think. A guy who turned in a stupid rating sheet might himself be faulted, for poor judgment.

I concluded that the best course was to play it straight. Shedding my personal likes, dislikes, and biases, I tried to rank my classmates based on their capabilities and potential as I saw them at that time. Did he plan ahead? Was he cool under pressure? Did he know the spacecraft? How would he handle the total responsibility?

After agonizing over the list for much of the night, I felt reasonably confident that I had been able to suppress my personal feelings. For example, I was able to rate Dave Scott next to the top even though there had been continuing friction between us. There was an edge to his needling that reflected, I thought, the disparity in our backgrounds. (Dave had the classic image: son of a general, in the top five of his West Point graduating class, hot test pilot. In Dave's view I was probably a smartass with the wrong pedigree: reservist, civilian, disrespecter of rank.) But his performance had pretty much resisted criticism and I gave him that.

At the same time I found myself writing in last the name of C. C. Williams, whose company I enjoyed, a fellow marine even. But C. C. lacked the academic credits of some of the others and, to keep pace, he struggled, cheerfully.

As I reread the list years later, I wondered how my judgment and that of the NASA brass (and the fickle finger of fate) could have been so different on Al Bean. I had placed him at the top in my peer rating. I thought he had it all: smart enough, the technical skills—he was a graduate engineer from the University of Texas—the experience, and the attitude. Yet he was the last in our group to fly in space. It's possible my personal feelings showed here.

To the outside world it must have appeared that everyone came out of the same factory. With a few exceptions, we were compact, short-haired, jut-jawed. And yet we made a fairly diverse cast of characters. The list below is in the same order as on the day I handed it to Deke Slayton. The capsule impressions were added later and some of those have changed over the years.

1. Alan Bean. Friendly, the prototypically cool naval aviator whose coolness, I learned later, was the result not so much of being on top of things, as of a refusal to let himself be ruffled. Both gymnasts by

training, Al and I hit it off from the beginning. Our birthdays were one day apart; March 15 (his) and March 16 (mine). We invited the Beans over for a joint celebration one year and discovered that before he ate out Al would check to see what his hosts were serving. He was the only astronaut whom Lo considered a more finicky eater than me. Al had a mania for spaghetti. There wasn't a city in the United States where we traveled that he couldn't tell you how to find a half-dozen Italian restaurants. With Al, it was a practical matter: a spaghetti dinner was one way of beating the expense-money pinch. (I would watch him dig his way through a plate of the stuff and remember the words of the old boxer Willie Pep: "Spaghetti killed more Italians than World War Two.")

2. Dave Scott. Despite our occasional personal friction, I recognized in Dave the controlled arrogance of those who are born to lead. He was made-in-America, big, strong, clean-cut, intelligent. He gave the impression that, though he might be wrong, he was never in doubt. Once, on a flight from Las Vegas to Prescott, Arizona, with me in the rear seat, Dave read the mileage numbers instead of the correct heading on his map and took off in the wrong direction. We were 50 miles off course when the Flight Control Center called to suggest to Dave that he ought to be correcting his route. It was characteristic of Scott that it didn't faze him and he never even let on that it embarrassed him.

3. Charlie Bassett. Another fair-haired, good-looking, All-American type. What most impressed you about Charlie was his self-discipline. On field trips he seldom packed a lunch, taking the position that it was good discipline to do without. While the rest of us were feeding our faces, he would use the time to take notes. He was wound tight, dedicated, had fine mechanical skills, and undoubtedly would have made an excellent space pilot—if he had lived.

4. Mike Collins. From a military family, he was the only foreign-born (Italy) in our group, a quiet, unassuming guy with a dogged disposition. Mike may have been the best-conditioned of the Apollo group. He was our handball champ from the first day we opened our new gym. Hard as nails, his physical fitness accounted, I suspect, for the success of his EVA mission on Gemini 10, where so many others failed. Mike and his wife, Pat, were our across-the-alley neighbors when we first arrived in Texas.

5. Bill Anders. A strong believer in the basic American ethic: hard work, fair play, clean living. Bill was among the more serious of our group, a quietly capable guy with a good academic background in nuclear engineering. He was one of the people with whom you could share a private thought. In retrospect, I may have rated Bill lower than

his credentials deserved, bending over backward not to favor a close friend.

6. Rusty Schweickart. The others probably judged Rusty and me on the same yardstick. We were considered the most natural competitors considering our similar backgrounds. They had automatically put the two civilians, the two research scholars, in the same office space. In truth, we were not that much alike. Rusty's very liberal views clashed often with my conservative ones. It led to interesting discussions, but neither could convert the other. We were intellectual friends more than emotional friends. A very open guy, Rusty was the one who could relate to the hippie, arty crowd and had friends who could admit, unblushingly, they had smoked pot. His wife Claire was an active campaigner for all liberal causes. At one time it was rumored she was the leader of a movement to blockbust the then lily-white neighborhood in which most of us lived, feeling with some passion that our kids were being denied the advantages of growing up in a racially mixed environment. Rusty was easily aroused by what he perceived to be social injustice. Of all of us in the program, he had the highest tolerance for his wife's independent activities.

7. Buzz Aldrin. From the outset, Buzz was referred to as "Dr. of Rendezvous" because of his obsession with rendezvous in space, the subject of his Ph.D. thesis at MIT. In fact, he looked more like a college professor than the fighter pilot he actually was. He smoked a pipe, eagerly engaged in "deep" discussions, and should have had leather patches on the elbows of his jacket. He was always intense and could talk for hours with anyone—as long as the subject was rendezvous. Once, when his wife Joan was out of town, we invited Buzz to dinner. Then I was called out of town and Lo just naturally assumed that Buzz would take a raincheck. Instead, at the appointed hour Buzz showed up, had a couple of cocktails and dinner for two, and then drank scotch until the wee hours, all the time giving Lo a monologue on the intricacies of rendezvous in space. Even at this early stage in our careers, Buzz struck some of the fellows as a bit odd.

8. Dick Gordon. There was a touch of devil-may-care in Gordon's bearing. Compact, tough, he had been an athlete of some renown at the University of Washington. He had test-flown the F-4, at the time the hottest thing in the military skies, long before any of the others had. His inside drag impressed the rest of us in the beginning. He knew half the guys in the space program by their first names, was sociable, well liked, and appealing to women. Our houses were a block apart, cost the same, were the same size and differed only in that the Gordons had three more bedrooms. They had six kids.

9. Gene Cernan. Another without a test-flying background. Gene's

only real strong points were his ability to get along with people, especially the right people, and his willingness to do what he was told. On the social circuit he was a charmer and would eventually surpass even Wally Schirra in this art. At the time of the ratings, none of us, including Gene, could have guessed that he would enjoy the most complete and successful space career in our group.

10. Roger Chaffee. In the early days, some tended to underestimate Roger, perhaps because of his small stature. But he had the capacity to fill a room—any room. It was impossible to attend a meeting with Roger and not be aware of his presence. He had a fighter pilot's attitude, even though his flying background was in multiengine photo reconnaissance aircraft. When confronted with a problem, Roger would bore right in. One of the youngest of the third group, he was fearless, confident, bright, with the All-American-boy look, and a beautiful wife to boot.

11. Donn Eisele. A hard one to figure. Donn had the proper test-pilot background and knew several of the guys before he came into the program, most usefully Tom Stafford. He seemed to enjoy the hard-partying and hard-driving aspects of our business as much as anyone, but I always had the feeling that he was pretending a little. He once confided that among the astros he didn't feel like much of a hard charger, but back in his old fighter squadron he was considered a real tiger. As a teenager in Columbus, Ohio, Donn had worked as a zoo-keeper. Both his parents died soon after he joined NASA, and one of his children, who was a source of much love, died of leukemia. Donn had more than his share of hardship to live with during our astro years.

12. C. C. Williams. I felt a twinge of guilt placing him so low, but long odds never bothered C. C. Marines were accustomed to getting by with less. He was an unspectacular guy, but his flying skills couldn't be faulted. He had performed all the automatic throttle tests for carrier landings with the F-8. He was a big, strapping six-footer who wouldn't let you dislike him. He had one other thing going for him: no family— or wife—to hamper his pursuit of the All-American fighter pilot image.

Ted Freeman, who is missing from the list, had been killed in an airplane crash only a short time before. He would have placed near the top.

To this day I have no indication of where Walter Cunningham stood on the other peer ratings. Certainly not in the upper bracket. I didn't figure to win the Mr. Congeniality Award. And I had to concede that there wasn't much evidence yet of what I could or could not do.

With some irony, a short time later, it looked as though it was all academic, anyway. I couldn't be concerned with where or how I rated because I was trying to keep the NASA doctors from washing me out of the program.

A really silly thing happened to me in February 1965: I broke my neck on a trampoline.

It happened at the official opening of the new astronaut gym at the Manned Spacecraft Center. Five minutes after the gym was dedicated I was working out on one of the two trampolines, under the watchful eye of a nineteen-year-old college gymnast. Along with Aldrin, Bean, and Carpenter, I was one of a handful of former gymnasts in the program who subscribed to the theory that trampoline exercises would enhance a person's adaptability to zero gravity in space.

After about ten minutes of working out, I decided to attempt what for me was a new trick, a "cody"—a front flip to a stomach drop followed by a back flip. At least, I think that's how you do it. If I had been able to get through it safely I suppose I'd remember.

The young fellow assigned to spot me (a spotter's job is to see that you're not hurt) watched as I proceeded to throw for and make the first cody I had ever attempted, although not very gracefully. Characteristically I bounced back up and told him I was sure I could improve on that first one. Well, it was a bust right from the beginning. I stalled out on the second half and came down right on my neck—fortunately, still on the canvas. I heard a strange crunching sound and a knifelike pain shot through my neck. My first reaction was that I'd have to spend the next day or so working the soreness out.

I asked the college champ why he hadn't spotted me and he replied, with some embarrassment, "I thought you could do it."

That was another early encounter with the "omnipotent astronaut" syndrome. In those days, astronauts seemed to stand about ten feet tall and people believed we could do almost anything. No doubt some of us thought we could. When I said I could do another cody, and a better one, he not only believed me, he stood back to get a clearer view.

He felt terrible when he saw I was hurt, and to save face—his and mine—I quickly attempted to shake it off. I moved over to another trampoline and spent about ten minutes in a game called "spaceball," but the pain was intense and I finally gave up and headed for the shower. Our gym, our brand-new gym, had been opened and dedicated at noon and a half hour later I was literally headed for the showers.

That afternoon, as I sat in class, it began to dawn on me that something was radically wrong. I could not turn my head. To look to either side it was necessary to shift my entire body, holding my head with both hands against the pain. I wasted no time in getting to the doctor's

office. From there I was sent to the hospital, where the diagnosis was
made: a broken neck, a compression fracture of the C-5 and C-6 verte-
brae.

The doctors assured me I had been lucky and would suffer no
paralysis. From that point on I wasn't concerned about the road to
recovery, only that the NASA flight surgeons and Deke Slayton might
leap to premature conclusions and write me off as physically unfit. As it
was, I'm sure many of my buddies were yielding to our customary
schizo reaction: "Poor Walt, tough break, but that's one less to worry
about for a flight."

There was only one way to fight back against the inevitable doubts
and suspicions and that was to prove from the outset that the injury
wouldn't handicap me. The moment my neck brace was fitted, I resolved not to miss a day of training, at the same time realizing I would
be grounded for an indefinite period.

Two days later I was in the front of the pack, hiking up the side of
Meteor Crater in Arizona during a geology field trip. I made damn sure
I was up front, more for the psychological effect on my friends (and
myself) than anything else.

The only real problem during the four months of living with the
neck brace was flying—or, rather, not flying. Flying is a lot like sex;
when you're getting a lot you always want more, if you have to do
without for a while you get more and more used to it.

Approval for flying came in easy stages. At first, I was limited to
flying co-pilot on the Grumman Gulfstream, a twelve-passenger executive aircraft. Ed White and Jim McDivitt were starting their Gemini 4
postflight activities and I became their chauffeur on trips to Washington, D.C., and their hometowns.

The doctors were even reluctant to let me fly as a passenger in the
T-38, because of the ever-present danger of ejection. They thought that
my already damaged back could suffer permanent disability if it had to
endure the effects of a rocket-seat ejection. During the third month I
was approved for helicopter flying. Only after I had shed the neck
brace—a month later—could I climb back into a T-38.

I never dwelt on the idea that the injury could have crippled me.
Although I shuddered at how close it came to wiping out so much
work, so many hopes, I came through with no ill effects. I was stronger
and more determined for having encountered adversity.

4

Getting in Line

BY JUNE 1965, even as we awaited the launch of Gemini 4 and America's first space walk, the old order was changing.

John Glenn had retired, explaining—with John's usual perception—that he did not wish to become the oldest, permanent training, *used* astronaut. Scott Carpenter and Gordon Cooper were losing favor. Al Shepard was grounded by an inner ear disorder, his future now uncertain, and Deke Slayton was buried under administrative paperwork. It was the hour of Gemini, with exotic promises of space rendezvous and docking and the sight of men outside the capsule, swimming in space. All of us Apollo jockeys were itching for a piece of the action.

Subtle changes were taking place and we were carried along as if on a moving sidewalk. Tensions were rising within the Manned Spacecraft Center. I wouldn't quite describe it as a power struggle, but it was apparent that the old NASA team spirit was taking a beating.

Up until Gemini 4, the astronauts would breeze into the various tracking stations around the globe prior to their mission. Always the last to arrive, they were quick to assume command. But now the star of Chris Kraft, the director of flight operations, was ascending, and he had different ideas on the relative roles of his flight controllers and the astronauts. Chris was emerging as a better organized and more powerful administrator than Deke, whose astronaut office had the glamor but was far harder to regulate.

49

In Deke's shop no one wanted a job limited to paperpushing. Not even Deke.

Chris Kraft had always wanted a flight controller in charge of each tracking site, and he eventually won his point. Our training schedule could really no longer stand the drain on the flight crew's time necessary for prime responsibility but the Astronaut Office reacted characteristically. If we couldn't be captain, we wouldn't play. We did, however, make one last evaluation of the new system during the Gemini 4 mission.

Bill Anders, Dave Scott, and I were assigned to man three of the stations. I drew the tracking site on the island of Kauai, Hawaii. Scott won a coin toss for Carnarvon, Australia, which meant running the gauntlet of the Sydney social attractions both coming and going, and Bill wound up with Guaymas, Mexico. We were advised that this was possibly the last time the astronauts would be used on such assignments.

That decision, it turned out, would depend in part on how we evaluated our roles. We were in the odd position of being able to cut our own throats on what was really choice duty by reporting that the tracking stations could get along nicely without us. Which they could and subsequently did.

But the three of us were cheered by the fact that someone had enough confidence to send us. I relished the opportunity to be on my own—away from the other twenty-nine, the only astronaut in town— and on the beach. It was a chance to flap my wings.

The flight to Hawaii was with Ed Fendell, a big, strapping bachelor flight controller, on his way to Carnarvon, Australia, to join Scott. There was a sort of gentlemen's agreement among those who manned Carnarvon that it would take at least four or five days—after the mission—to close up shop. The boys never found it easy to tear themselves away. The final reports were always late arriving from Australia, and for some unaccountable reason some of the fellows would get lost in Sydney and miss their airplane connections home. After John Glenn's historic flight, they had to send a telegram to the American Embassy to locate Gordon Cooper—apparently lost in the wilds of Sydney.

Needless to say, Ed Fendell was in high spirits as our Pan Am flight droned over the Pacific Ocean. Ed and another controller were partying it up, flirting with the stewardesses, and having a great time. The other guy was playing his guitar when Ed reached into his briefcase and fished out a stack of eight-by-ten photos, mostly of John Glenn.

"What's all that for?" I asked.

"Trading material," Ed answered. "You'd be amazed what I can get in Australia for one of these pictures. How about signing a few of yourself?"

He studied one of the prints and said gratefully, "John Glenn has no idea how many times his picture has gotten me laid."

I laughed out loud. "Hell, Ed, I'm not John Glenn. Nobody knows me."

"Doesn't matter," he assured me. "You're an astronaut. Won't make a lick of difference." I had yet to learn the facts of life as Ed took them for granted.

I checked into the Kauai Surf Hotel on the beach, and Lo flew over a week later—at my expense—to spend ten days. It was her first chance to travel with me and share some of my involvement in the astronaut business. It was the perfect spot, all sunshine and moonlight, orchids on your breakfast tray, swaying palms and the scent of fresh pineapple, and at night falling asleep to the rhythm of the surf against the shore.

Our hotel was a favorite retreat for movie figures and assorted celebrities. Raymond Massey and his wife moved in a few days after we did. He checked out the next morning, complaining that the surf kept him awake.

But those with a little romance in their souls found no fault with it. For most of the astronauts those rare excursions were a blessing and a tonic to marriages that were under continuing strain.

From where we sat, Gemini 4, commanded by Jim McDivitt, went off smoothly. It was from our tracking station that the word went out that Ed White was "Go for EVA," the first American space walk. After twenty minutes, we had a helluva time getting him to go back in.

Chris' man was in command. I went along, avoided rocking the boat, and operated the retro-fire and computer clocks. Later, with great reluctance and a sense of loss for the astronauts yet to come, Dave, Bill, and I reported to Deke that the tracking stations could manage just fine without us.

Several months had passed when Deke called a meeting of our Apollo group during a geology field trip in Oregon. We slouched into chairs and took up key positions on the floor of Deke's motel room, waiting for what we sensed would be important news. It was not unlike the scene in which the colonel gives the old window-shade yank to the wall map and announces to a breathless squadron that the long wait is over.

In his no-nonsense, to-the-point way, Deke opened by saying that some of us were going to be moved into the Gemini flight rotation. Initially, NASA had thought that the first two groups would be able to handle the flights through the Gemini program. But the flight schedule was becoming increasingly hectic, with launches every ten weeks, sometimes even closer, and the Apollo program would soon require the assignment of three-man flight crews.

I guess the last point was as significant as any. The Apollo development was being held to schedule while the Gemini program was nearly a year late.

In a matter-of-fact manner, Deke announced that Mike Collins would be backing up Gemini 7 (with John Young), Dave Scott flying Gemini 8 (with Neil Armstrong) with Dick Gordon backing him up (with Pete Conrad). The significance of this did not go unfelt. Dave was flying and Dick and Mike were *in line*. At the same time Mike's future had been hitched to Frank Borman's star and Dick was set for a long ride on Pete Conrad's coattails. They had clear shots at the downstream flights.

In the next few days they almost magically disappeared from our routine. We wouldn't see them for weeks at a time. They were playing off-Broadway, while we were still in Bridgeport.

We would try to catch one of them in the office halls for a moment, long enough for a quick question about what the big boys were doing. Their lives had changed, and the rest of us marveled at how easily and abruptly it happened.

Among those of us still waiting, the attitude was, "Well, if they're in line, I can't be far behind." There really wasn't time to feel any pangs of envy. Much.

As the weeks passed with more training, more classes, and virtually no new information our frustration began to build. In October 1965 the second anniversary of my selection as an astronaut came and went, and I felt as though I were trapped in zero gravity in the middle of a large tank and couldn't touch the sides. I was wondering when someone would throw me a rope. No one had yet said a word to me about when, or if, I'd ever make a flight.

There was a brief respite in November as I returned to Kauai for four days. The occasion was the first Conference on Oceanography and Astronautics. Oceanography, then the rage, was receiving more and more government funding, much as the environment and ecology did a few years later. Gov. John Burns of Hawaii was attempting to attract private and government oceanography and aerospace investments to his state. Dr. Hugh Dryden, deputy administrator of NASA, had been invited to give a luncheon address and to bring along an astronaut who would present one of the papers and be a potted-palm embellishment.

Normally we stayed in one place only long enough to finish our business; then it was on to the next commitment. This, however, would be a four-day visit with my only obligation a one-hour paper to be presented on the second day. This left me with plenty of time to lie on the beach and study Gemini engineering documents.

Very little studying was actually accomplished because the first person I ran into on the beach was Clare Boothe Luce. She was then sixty-two and well deserving of the reputation for charm and intellect which she had gained over the years. She had already had successful careers as an actress, writer, editor, war correspondent, playwright, congresswoman, and ambassador, and was at that time one of the leading advocates of the relatively new science of oceanography. Her presentation addressed the problem of getting more funding for the emerging and essential science of oceanography. We spent the next couple of days on the beach discussing the problems of government funding for an emerging science.

From time to time, we would take a dip in the bay but never let it interfere with the conversation. That nearly got us into trouble.

On one occasion we were idly swimming toward the mouth of the bay, chatting as usual, and forgot the time. She swam along very slowly in an easy little breast stroke, while I did a backstroke alongside. When we finally stopped to take our bearings, we were at the mouth of the bay and the open sea. It was definitely time to start back. The three-quarters of a mile outbound had gone fairly quickly because we had been swimming with the tide. The way back looked like forty-five minutes to an hour against the tide. I was sticking close and trying not to show my concern for her, but she just kept chugging along.

After another fifteen minutes we noticed an outrigger canoe coming toward us from the beach, and when we finally rendezvoused, it was apparent the lifeguard onboard was concerned for our safety. Explaining they had lost many swimmers in just this fashion, he began to escort us back to the beach. I expected Mrs. Luce to get into the canoe, but that just wasn't her nature and she insisted on swimming all the way to the beach. She may have finished fresher than me, even though I was spotting her nearly thirty years. With all her other accomplishments, Clare Boothe Luce is one hell of a swimmer.

It isn't in my nature to passively sit by and let the fates play on. By early December 1965 I could no longer contain myself. The office I shared with Rusty was next door to the one occupied by Schirra, then in charge of our training group. One day I just walked in and unloaded.

"Wally," I blurted, "what's the story? I've worked my tail off for two years and here I am still sitting on the bench. Is that what's in store for the rest of my career? Warming the bench?"

Wally hesitated a minute. "Take it easy, Walt, you're on the first team. I wish I was free to tell you more right now, but I can't. Don't worry about it. They'll let you know before long." If Wally wasn't part of "they," who in the hell was?

He had really told me nothing, but I walked out of his office feeling elated. In my anxiety I had interpreted Wally's mild reassurance to mean my turn was coming.

In January 1966, the word came down that Gus Grissom, Ed White, and Roger Chaffee had drawn the first Apollo flight. It was a nice plum for Gus, the second American in space and the first to command a Gemini mission. The milestones of a test pilot's career are the rare first flights of a brand-new aircraft (or spacecraft), and now Gus would have logged two in a little over two years.

A few days later I was called into Schirra's office along with Donn Eisele. With a benevolent grin Wally announced that the three of us were to be the prime crew for the second manned Apollo flight. I was excited and at the same time a little disappointed. We would be flying a mission which was a virtual repeat of the first Apollo flight and in a virtually identical Block I spacecraft. So many approved changes had piled up that beginning with the third spacecraft on the line, the Command Modules had been redesignated Block II. The Block II vehicle was much improved and we had even begun to question the capability of the Block I vehicle to perform any of the design missions. There was some office gossip that the Block I missions were just to keep the program moving while we waited for Block II to come down the pike. The gossip would prove to be well-founded. No Block I spacecraft would ever get off the pad.

Flying virtually identical missions in outmoded Command Modules, when the Block II vehicle was so much better for the program, was not my idea of the planning and efficiency that NASA had been noted for. Consequently, Donn and I never really felt comfortable about the permanence of our mission in the flight schedule. But our logical analysis of the situation was swept aside by the elation we felt and we went out to celebrate. We were glad to have any mission. The answer to why it was there at all wouldn't be known to me for another eighteen months.

There has always been a general impression—not just outside the program but among some of the NASA people as well—that flight crew selection involved a painstaking, almost scientific, process. It was pictured as not unlike a computer dating service where they fed punch cards into a box and, after a lot of buzzing and whirring and blinking of lights, out pop the names of two or three people with complementary strengths and personalities.

The selection of Wally, Donn, and myself, if you listened to the news media, reinforced this theory. On the surface we were compatible: Schirra was the old master, experienced, with an easygoing, gregarious personality; I was depicted as the outspoken, intense, hard-charging, self-made young guy; and Eisele registered about halfway in between.

(A year later, Donn confided that he had originally been ticketed for Gus Grissom's crew but had been dropped when he developed a shoulder problem that required an operation. Roger Chaffee replaced him and Donn was reassigned to the next flight. I also found out later that Wally was originally assigned as a stand in on Apollo 2 and Donn was a replacement. God only knows if I was originally ticketed for that mission.)

At that moment life was too sweet to nag about the system. My future seemed to stretch ahead of me like a dragstrip to the stars. I was on the flight schedule ahead of most of the fellows in my group and had completely skipped the backup-crew phase. With launches scheduled every eight weeks, it looked like a fat space career was ahead.

The down side was the strong suspicion that I was out of the running for the first lunar landing. In the natural order of things, flying the second Apollo mission would keep one off another prime crew until at least the eighth launch. And we planned to go to the moon before then.

My personal objectives were threefold. The first lunar landing would have been icing on the cake—too much to really hope for. First, I wanted to get into space, which now seemed imminent. Second, I wanted to make a lunar landing. There figured to be as many as eight or ten, meaning that at least sixteen of us—out of the thirty in training —might walk on the surface of the moon. I reckoned the odds at 50–50 and that didn't seem bad. Third, I wanted to command my own mission.

To fly in space, to reach the moon, and to command a mission would mean maturing as a man, would satisfy a need to test myself fully and—overlooked by none of us—would not exactly be a handicap in the commercial marketplace.

Unlike some, I had not joined the program expecting to stay forever. These goals, if I could achieve them, would be good stepping-off points. By now we had before us the example of John Glenn, who had signed as an executive with Royal Crown Cola for $50,000 a year and stock options. While we all looked on John as the first great space age hero and knew we couldn't count on scoring so big ourselves, we reasoned that the technical achievements still to come in Apollo would lead to similar opportunities. Standing at the gateway to the moon, Glenn's three orbits in that simple Mercury capsule no longer seemed awesome.

The assignments were now complete for the first two Apollo missions. The Grissom crew would be backed up by Jim McDivitt, Dave Scott, and Rusty Schweickart. Our backups were Frank Borman, Tom

Stafford, who showed up in May, and Mike Collins, who joined us for about two months starting in August. Tom and Mike were tied up with the tail end of the Gemini program.

The twelve of us were now joined like lab mice to get the program off the ground. Overnight we found ourselves working on a more personal and specific basis with the prime contractor, North American Rockwell. In the role of engineering and operational consultants, we would be spending most of 1966 checking out hardware and developing test procedures.

Our relationship with North American Rockwell was, from the beginning, a curious one. It was their first effort in manned spacecraft and most of the knowledge and actual experience in manned spacecraft design and fabrication was residing at the McDonnell Aircraft Company in St. Louis, Missouri. MAC was where they wrote the book. They had created both the Mercury and Gemini spacecraft. North American Rockwell was on deck with little, if any, experience but a whole new generation of technology.

Putting their best engineers and scientists on the proposal for a vehicle to carry men to the moon, North American Rockwell had won out over tough competition. The Space Division was headed by Harrison (Stormy) Storms, who had previously ramrodded the X-15 rocket research plane program.

NAR was moving into a strange new world of manned spacecraft. Their engineering and technology was a step more advanced than McDonnell's on Mercury and Gemini, but also more troublesome. One development problem surfaced after another. They were continually fighting costs and schedules as they seemed to be trapped in that infamous cycle, "The hurrieder I go, the behinder I get."

In their eyes we must have seemed like spoiled brats whose whims had to be tolerated. We were into everything, not just the controls and displays and the crew couches. Was the environmental control system engineered to do what we felt would be demanded of it? Were the records for test and checkout procedures adequate? Even: were the technicians dressed correctly for the job?

The astros who had worked with McDonnell were quick to observe that North American Rockwell was stumped by procedural and development problems which McDonnell's experienced crew could have handled routinely. It took the rest of us somewhat longer to reach the conclusion that North American Rockwell was pretty screwed up, and that the spacecraft they were building was sadly lacking.

We debated the potential changes amongst ourselves and usually agreed on which ones should be pursued. But no doubt about it, individual instincts—and hangups—did enter into it.

Roger Chaffee, for example, had what we referred to as a micro-design craze. Roger was never content with just saying that the communication system should be changed to meet such-and-such a requirement. If there was a deficiency he would sit down and, in four or five minutes, design his "solution." The danger in this was the perfect "out" it provided the contractor when his "back of the envelope" solution did not work. "Well we did it like Roger said," was an excuse heard more than once in the last half of 1966.

Wally rarely argued theory or design, preferring to wait until the system was solid hardware. It was at that point that he would suggest whatever adjustments he felt were necessary. The impacts were much greater, and the costs much higher. He often resorted to trading Brownie points or a power play to accomplish things. Wally was the kind of guy who felt, intuitively, how he wanted a system changed. If necessary, he would turn the issue into an emotional vendetta. Fortunately, Wally's intuitions were often the right ones.

But between NASA's demands and Rockwell's learning processes, the pressure was growing on both engineers and managers. The astros had no legitimate program authority. We were engineers and managers without portfolio, so to speak. We had the glamor and the glory on our side because we were the ultimate users. In effect, it was our parachute they were packing. Our de facto authority came from this role as operator.

Even when the contractor disagreed with an astronaut about changing a bolt, he was aware he was talking to the man whose life might hang on that bolt. If he forgot, some of our guys weren't bashful about reminding him. We were concerned with just two questions: was it safe and would the spacecraft perform what it had to? The contractors had these and other monkeys on their back. Such as cost. And schedules.

At one point Wally insisted that the Service Module engine installed in the flight vehicle be test-fired on the ground. But once the engine was fired it had to be completely rebuilt, meaning that no matter how many times it was tested the subsequent rebuild would leave it a virgin. But Wally wanted to be shown that the engine worked. He pursued his case at the executive offices of North American Rockwell, at several levels of NASA, and—most effectively—at parties and other social functions. Sure enough, we test-fired the engine. At the cost of a few million tax dollars Wally felt vindicated, the contractors were relieved to get him off their backs, and to this day no one can say if the test proved a thing. When Wally got involved in emotional causes it was usually late in the game, without wasting a whole lot of time on his homework.

Over the years the astronauts had acquired a mixed reputation among the contractors. They came on board with engineering skills and technical abilities that ran the gamut from good to don't-ask. The Mercury guys were basically test pilots, not exactly enthralled with engineering details. The Gemini group not only had solid test-pilot credentials, but also averaged nearly five years of college. Some of them were fine engineers, and they made significant contributions to the Gemini design. Our Apollo group stood lower than the first two in flying credits but was the heaviest in academics.

When our crews arrived on the scene, we crawled over everything like a pride of lion cubs. It soon became obvious we could do great good or, without really trying, cause serious harm. Some of the engineers at Rockwell were so eager to please that the merest hint from an astronaut would be picked up and quickly run through an entire design concept.

On the other hand, an astro with a valid idea might have to throw a tantrum at upper management levels to get it into the spacecraft. Many times we would have to overcome the resistance created by earlier astro changes—some of which were unwise, some even capricious.

At a two-day meeting in Houston attended by the top brass of both NASA and North American Rockwell, the credibility problem broke into the open. It was beginning to appear that cost and schedule over-runs on the Command Module were going to be, pardon the expression, astronomical. The meeting was chaired by Dr. Robert Gilruth, the director of the Manned Spacecraft Center. At one point a management type from Rockwell stood up and said, "Look, we've got one problem after another, but we could be doing all right. We were on schedule and moving along just fine until the astronauts came out there and began working with us full-time. Now it's going slower and slower and slower."

Dr. Gilruth did not take kindly to outside criticism of America's space idols. He still trusted our judgment. He fixed the old boy with a cool stare and said, "That's OK. We're not charging you for their time."

All joking aside, we were really needed, and much of the work we did there was crucial. If anything, our impact was felt too little, not too much. The problems eventually were corrected, but it took a fire and an additional two years to do it.

Those were strange and overlapping days. We were living with brand-new machines, the only two Block I spacecraft in existence. Gus' crew was running several weeks ahead of us on the checkout, and they bore the brunt of creating and revising test procedures. Testing went on around the clock and yet we kept falling further behind. Along with our work the social temptations were murder. Everyone wanted to wine and dine us. There was one party invitation after another. The prettiest girls in the plant seemed to have very big eyes, and the big-stud–astro image required that we stare back.

We literally lived at North American Rockwell from Monday morning on through the week, usually flying back to Houston on Friday night or Saturday morning. As much as we might have been tempted, it was impossible to keep up the work pace and still play the bon vivant. Anyone who really tried would have ended up in a jar in the Harvard Medical Laboratory within six weeks.

Frank Borman and I pulled the midnight checkout duty about 75 percent of the time, with Mike Collins doing his share during his brief association with us. You learn a good deal about the thresholds of men in tedious, demanding situations. Gus Grissom, for example, would take his whacks at the good and the bad along with everyone else. He was a hard liver and loved to party, but if Roger Chaffee, the youngest astronaut in the program, was pulling some notably boring duty, you were likely to find Gus sharing it with him.

Although we made some of their engineers nervous, the customer relations types at North American Rockwell could look upon us as a kind of tourist attraction. Our group included some of the early glamor boys—Gus, Wally, Borman, Stafford, White, and McDivitt—who had been in space and whose names were known to the world. Also, our crews would be the *first* to fly the *first* Rockwell spacecraft.

We were welcomed into the executive dining room, which provided the perfect forum: it was a free lunch and a chance to sell our ideas at the same time. It was here over lunch where we would socialize with the top executives, lobbying for whatever changes seemed urgent, maybe even undercutting some engineer who was trying to put the brakes on what we wanted done. By the time he got to a meeting where it was discussed, his boss would already be convinced that it should be done our way. That may not have been in the spirit of fair play, but it was effective.

Since the dining room seated only thirty people, with our force of twelve we could practically flood the place. Sensing a crisis, we got together and decided that we would restrict ourselves to no more than five at one time. Even then we had to exercise discretion on which five. Ed White and Dave Scott, for example, could eat their own weight in shrimp cocktail. Either of them could put away a dessert that any self-respecting horse would choke on. And Jim McDivitt could scarf it down with the best of them.

In the process of all this work at Rockwell we became a whole team. Behind the prime and backup crews we had a support crew and a

dozen NASA engineers who worked with us on the systems, the procedures, and the on-board data. They all became a part of the Spacecraft
014 team. It was *our* spacecraft.

Manned space missions have always had a prime and backup crew
assigned. Because of the increased complexity on Apollo we added a
third crew. Deke's intention for the support crew was precisely that:
support, not training. Their real responsibility was to free the prime
and backup crews so they could spend a greater part of their time on
productive training. There were only so many meetings, reviews, tests,
and simulations in which we could participate. In addition, many of
the scheduled activities had a tendency to be inconclusive or leave the
impression that the exercise was unimportant if an astronaut was not in
attendance. Support crew members provided that presence. Many
times one of them became the crew expert in some essential area of the
mission. Predictably, they also inherited any shit details which came
our way. In a world where the newest astros, at the bottom of the
pecking order, were looking for any handhold to climb up, or any
opportunity to come to Deke's attention, a support crew assignment
was looked upon as a real plum.

In eight years I spent a total of ten weeks on a backup crew between
prime assignments. But few got off so lightly. A more typical pattern
was to progress through support team, backup crew, and then prime
crew. That process could take years. This third crew was invaluable to
the test and checkout phases, and in time, that assignment became a de
facto prerequisite for crew assignment. Deke just naturally gave more
consideration to those who came to his attention by doing a good job
in this rather thankless assignment. There was no rookie crew member,
from Apollo 13 on, who did not spend at least one tour on a support
crew.

Most of the capsule communicators (CapCom) eventually came to
be drawn from the support crew ranks. The CapCom is a key control
over the quality and accuracy of information transmitted to the crew in
orbit. Not so subtly, he also helped maintain the fiction that *no one*
tells the flight crew what to do except another astronaut. There was an
astronaut support crew member in Moscow during the joint U.S./Rus-
sian mission in 1975, in the event direct communications were required
with our spacecraft.

From the coaching staff, Slayton or Shepard would normally represent us at upper management meetings. They defended the flight
crew's position at the many review boards at North American Rockwell
and NASA.

For major, week-long reviews there might be several hundred participants divided into teams addressing the various subsystems and

operations of the spacecraft (the guidance and navigation system, the electrical system, quality control documentation, test procedures, etc.). The Astronaut Office was directly accountable for the Crew Station and, in addition, we had representatives on most of the other teams. Deficiencies or problems in any area were pursued from our role as user and pilot of the vehicle.

On Wednesday night, after three days of grinding it out on the teams, the flight crew would caucus to make priority judgments on those points we wished to pursue. Our lists in the early days of Apollo ran into the hundreds. On Thursday a screening board with contractor and NASA representatives disposed of those items on which they could agree. On Friday the official Review Board convened to endorse the activities of the preceding four days and disposition those items which were irreconcilable except at the highest levels of management. This board would be co-chaired by Dr. Gilruth and the contractor's program manager. Chris Kraft would represent flight operations and Deke Slayton would represent the flight crew.

Review weeks were tough with long hours, but it was the formal way of shaping the spacecraft to eventually fly the missions for which they were intended.

Wally was not only our spacecraft commander, but also the captain of the team. We all had a fairly free hand when it came to scheduling our activities, but if Wally said, "Get your fanny to the Cape," that's where we got our fannies.

We laughed at the Jet Set and traveled so frequently and easily that we never had reason to become bored by the immediate scenery. It was easy, if we began to feel frustrated, to pick up and fly off to a different duty. We had so many legitimate reasons to move around the country that a "business trip" to some particular location could be arranged very quickly if we felt some personal urgency to be there. It was just as easy for official business commitments to be delayed or to vanish completely if there was no T-38 available. Who needed a trip by commercial airliner? Talk about prima donnas, trips were sometimes put off simply because airplane availability made it necessary for us to ride with another pilot. Who wanted to be a backseat passenger half of the time? We were literally spread to the winds—but, I should add, generally in the interest of the mission.

Take the morning Lo and I rose, as usual, at 7:00 A.M. Almost immediately she took a call from Al Shepard. Handing me the phone, she commented, "You've got to go to Michoud this morning."

Al had received a call only a few minutes before and was passing on the details of a command performance. "Walt, President Johnson is going to be in Louisiana this morning, touring the Chrysler plant at

Michoud. James Webb wants the Apollo Seven crew to go around with him." Webb, the NASA administrator, was a master politician and a respected associate of the president. He recognized and fully exploited the value of publicity to the space program as well as the glamorous effect of having "his boys" escort the president.

"The president," Al was saying, "and James Webb and Governor John McKiethan are scheduled to be at the plant at ten fifteen this morning. We may not be able to get Donn there in time [Eisele was at MIT studying the Apollo computer], but I'm calling Wally at the Cape now and he'll meet you there."

Such short notice was not unusual, but the reason was rarely as auspicious as an appointment with the president—my first with any president. I scheduled a T-38, gulped breakfast, grabbed a good suit, and headed for New Orleans. It's a short flight from Houston but nearly an hour's drive to Michoud from the Naval Air Station, so there wasn't much time. En route I arranged for a navy sedan to pick me up on arrival and tried to determine Wally's expected arrival time. After landing and changing, I delayed as long as I dared before heading for the Michoud plant. You just don't keep the president waiting. I arranged for a second car to wait for Wally, but just as I was pulling away he breezed into a landing and we drove over together.

The tour was meant to inform the president about the Saturn boosters which were stored at Michoud, and, at the same time, to be a morale booster and publicity trip for the working troops. We were in a recovery phase after the pad fire, and the program needed a shot in the arm. It was a routine space industry tour, but three things stand out in my mind. One was the briefing to Wally and me from Jim Webb: "OK now, you two get right in there every time he stops, you get right up by the president and see to it that you're in the pictures. Don't be bashful and don't wait for someone to tell you to get there—you just be there." Jim meant what he said, to the point of hanging back himself whenever possible and continually urging Wally and me to "move in among them."

At Webb's insistence, Wally and I joined the president in his limousine for the short drive to the front of the plant. President Johnson relaxed and made casual conversation.

"Jim Webb is the only man working for me," LBJ declared, "who told me what he was going to do, when he was going to do it, and what it was going to cost, and is accomplishing all three." He was referring, of course, to the lunar landing program.

A few minutes later, just before the short trip was over, the president observed, "It's unfortunate, but the way the American people are, now that they have developed all of this capability, instead of taking advantage of it, they'll probably just piss it all away." That 1967 statement is proving to have been quite prophetic.

This interlude lasted but a few hours—Wally headed back to the Cape and I spent the afternoon working in Houston. Late that night, I flew on to the Cape and the next morning had breakfast with Wally. We both commented on how casually we adapt to what is really a fantasy-like existence: our business day had been interrupted by a two-hour interlude with the president of the United States, with no warning and essentially no afterthoughts, and we were back on our old routine again.

As if to assure that we would not lack for flight hours, the various simulators connected with the mission were scattered all over the country. These were machines or instruments designed to simulate the functions of the hardware or other aspects of the mission. The simulators were being modified and updated on a continuing basis, so of necessity many were kept close to the contractor.

Some were fairly simple in concept. At Chapel Hill, North Carolina, and at the Griffith Park Observatory in Los Angeles, California, we did our planetarium work. We simulated looking at the stars either through the guidance and navigation sextant, or through the telescope eyepiece. The object was to become expert at identifying the thirty-seven Apollo navigation stars as well as the hundreds of other twinkling signposts used for finding them. The trick was in doing it while looking at only one twentieth of the sky, the view we got through the Apollo telescope.

You can get the same effect by looking through a toilet paper roll, shortened to approximately three inches, which is what we did to simulate it as we traveled from coast to coast, mostly at night.

When performing navigation tasks, or "taking a fix in space," one identified an "Apollo star," placed the cross-hairs of the on-board sextant on it, and pushed a button which fed the information to a computer. Marking on three Apollo stars permitted the computer to determine one's attitude and position in space.

It had become obvious early in the astronaut game that it would be impossible for us to meet our schedules without our own fleet of jets. When I reported to NASA in 1963, eight T-33's—the oldest operating jet in service—were available, along with four of the newer F-102's. At that time NASA was evaluating other aircraft for "astronaut flight proficiency training." The Flight Operations Division selected the T-38, a two-place supersonic training aircraft, which was just coming into widespread use in the air force.

There were few hotter airplanes going at that time. The T-38 had the best high-altitude cruise speed, and when it was light on fuel, it was possible to climb from the ground to 40,000 feet in less than two minutes. It was safe, reliable, and enabled us to maintain the man-killing schedules that would otherwise have been intolerable.

With a T-38, any day in the life of an astronaut could be slightly supersonic. Many mornings I would get up at 6:00 A.M., have a quick bite with Lo, drive eight miles to Ellington Air Force Base, and be airborne by 7:00 A.M. The flight to Los Angeles required a fueling stop at El Paso, Texas, but because of the enthusiastic support of the ground crew there and the two-hour time change heading west, it was possible to land at Los Angeles International Airport shortly after 8:00 A.M. As I parked the plane, a North American Rockwell helicopter, arranged for the day before, would be dropping out of the sky to pick me up. It was only a ten-minute chopper ride to the space division plant at Downey, and with time to spare, I'd breeze into an 8:30 meeting with James Bond coolness.

After eight hours of technical talk and a business lunch at the golden trough (our nickname for the executive dining room—it never ran out of food), I'd catch the last helicopter back to the airport and ten minutes later be airborne in the T-38. By 5:30 P.M. the plane would be cruising at 41,000 feet and I'd be relaxed and pulling the paperwork out of my briefcase to get as much work done as possible before the sun set. Since we had no autopilot, it took both experience and clever work with your knees to keep the plane on an even keel. Even allowing for the two hours we lost heading east, with any luck and a good jetstream I'd walk in the front door at 10:00 P.M., have a nightcap with Lo, and hit the sack—maybe to do it all over again the next day.

Without these expensive toys—in some ways they served as airborne pacifiers—many of us would have gone bananas trying to maintain our schedules. After a day with slide rules and hardware, we looked forward to the peace and quiet of a night flight between Los Angeles and Houston. It was the finest airplane I have ever flown and I never tired of it. After seven years, I would still take off with both afterburners lit, pull the nose up to watch the ground fade away, and exclaim, quite involuntarily, "Jesus Christ, what an airplane."

Fun and Games

IT WAS a dizzying pace, but we thrived on it. It was a living demonstration, at least to ourselves, that we led glamorous and exciting lives that few others would be able to handle. Naturally, the price for this accelerated life had to be paid by the other half of the family.

We were busy tempting the Fates and playing hero, and our absences grew so frequent that our families came to expect them. Almost routinely, family responsibilities were being passed on to our wives: balancing the checkbook, seeing that the lawn was mowed regularly in the summer, renewing a bank loan, overseeing house construction.

Many of the domestic problems some astronauts faced later were due to our abdicating responsibility for so many years. I tried to cope with it by carrying on as though I hadn't been gone at all. The kids learned to accept it. When they were bad and deserved discipline they got it when I returned if Lo hadn't already punished them.

In our family the problem was never acute. The adjustment the kids made to the demands of my job often amazed me. Once, while I was gone, Lo overheard them playing house.

Kimberly said, "I'll be Mommy," and of course Brian was Daddy.

After a few preliminaries it was time for Brian to leave the house, as daddies do, for his office. "OK," he said, "I'm going to work now. See you in a week."

Lo almost dropped the dishes when she heard that.

One effect of our constant travel was that our secretaries wound up acting as liaison with our wives. We were gone so much we couldn't afford to call home on any regular basis, and we weren't allowed to use our government credit card for personal matters. So we learned to relay and receive messages through the office. Once my secretary, Charlotte, called me at the McDonnell plant and passed on a message from Lo telling me the refrigerator was broken and it would cost $125 to repair. The unit was eight years old and, according to the repairman, wasn't worth fixing.

I thought a moment and said, "Tell Lo to call Charlie at General Electric and see what kind of a price he can get us on a new one." Charlie was not an appliance salesman; he was the GE representative to NASA, a space guy. But in a pinch we seldom hesitated to call on our contacts.

The next day I called the office. Toward the end of the conversation I asked Charlotte how Lo made out with the refrigerator. She said she'd check and call me back.

That afternoon I got another call from the office: "Lo says the new refrigerator arrived."

As for our neighbors, we might as well have been in the CIA. They came to think of each of us as the Shadow, so seldom were we caught at home. Slowly we drifted apart from all but our closest friends. Those friendships we did form were not cemented the way friendships are when people share their social and domestic pleasures. When home we tended to hibernate, to stay close to the hearth.

We had grown jealous of our privacy, so we clung to our families as an escape from the limelight and the exotic life we led when we traveled. Having established, while traveling, lifestyles that included dinner parties, celebrities, the rich and the famous, and people seeking our autographs, we became aware of the sharp contrast to the lifestyles we knew our wives had at home: nothing socially, the washing machine on the fritz, the kids misbehaving, the loneliness. When the space voyager came home, all he wanted was to sink into an easy chair and watch television. After being exposed to a wide variety of the human species, including the female half, we were able to appreciate even more the strengths, the abilities, the known qualities of the women who kept our homes. The irony of it, though, was being unable to convince them of it. Our wives were growing more and more insecure.

Through the same media treatment all of us were getting, our wives came across as grown-up drum majorettes whose ambitions in life had been to play the piano and bake cookies. The size of our egos was exceeded only by their sense of insecurity.

While many of us were struggling to get the Apollo program out of the starting blocks, the Gemini program was finishing up a two-year action-packed flight program. It was the proving ground for nearly every critical step of the space program for the next ten years. Although we were still feeling our way, the Gemini spacecraft was not the experimental laboratory model of the Mercury program. It was produced on the nearest thing to an assembly line for space hardware we had seen so far, lifting off from Cape Kennedy at regular two-month intervals, and performing just fine. The program had been conceived as a gap filler between Mercury and Apollo, and it is now hard to imagine going to the moon without it. Gemini's contribution was purchased at the bargain-basement price of $1.5 billion.

The entire system, from the checkout crews to the flight controllers and the worldwide tracking network, learned how to gear up routinely for an operation that would never be quite routine. Gemini made three essential contributions to the development of manned space flight and the eventual success of Apollo.

The first was in June 1965, when Ed White stepped out from his tight but secure little cabin on Gemini 4 and into the cold vacuum of space for the U.S.'s first extravehicular activity (EVA). Ed's walk in space came nearly three months after Aleksei Leonov, the Russian cosmonaut, had become the first man to venture outside a spacecraft cabin into the airless void of space. Ed's fifteen minutes outside was little more than a feasibility demonstration or, more accurately, a dramatic stunt. But the same thing could be said of Leonov's. The Russians have not developed their capability much beyond the stunt stage, while EVA was utilized and improved upon in half the Gemini missions. It was here where we cut our teeth on a capability that has been an integral part of all our space operations since 1966.

Neither the Apollo-Soyuz mission nor the Skylab program would have been possible if we had not learned how to routinely rendezvous with orbiting spacecraft—the second essential contribution of Gemini. We would not have attempted a single lunar landing without certain knowledge that the Lunar Module with two crewmen on board could rendezvous with the orbiting Command Module for return to earth. The technique may have been improved a bit on the early Apollo missions, but all phases of rendezvous were developed, tried, tried again, and perfected on the last six Gemini missions. It was on those everyday missions, with launches every two months for two years, that the studies and trade-offs were made with navigation schemes, tracking techniques, spacecraft control modes, and so on. Rendezvous became a routine operational technique that now enables us to join up with just about anything only hours after lift-off.

The Gemini spacecraft cabin contained seventy-five cubic feet of what was inappropriately described as "free volume." That is roughly equivalent to the front seat of a small foreign sports car. It was anything but free after filling it with essential equipment and two human bodies dressed in spacesuits. Even the Mercury spacecraft had fifty cubic feet of free volume for only one warm-body intruder.

It is difficult for me to imagine myself cooped up in such tight confinement for even forty-eight hours. And that may account for my admiration of the premier human accomplishment of the entire Gemini series. Time is relative, but I think most would agree that the first "long duration" mission was that of Gordon Cooper and Pete Conrad on Gemini 5. Their motto was "eight days or bust." Doctors, both inside and outside NASA, had expressed doubt that man could adjust to life in zero gravity. Some went so far as to predict that exposure for a long period would probably be fatal. The astronauts viewed such pessimism as a "straw man," created to enlarge the participation of medical doctors. The opposing views would soon be put to the test.

In December 1965, Frank Borman and Jim Lovell climbed out of their Gemini 7 spacecraft on the deck of the aircraft carrier *Wasp* after spending fourteen days in those unbelievably cramped quarters. That was longer than any of the planned Apollo missions and would hold the U.S. duration record for the next eight years. It was a very human triumph. Of course, the equipment had to work to sustain them, but machinery can be designed to operate continuously for a long time, and its performance is not complicated by feelings. Yet at the end, the equipment was in much poorer shape than the passengers. The crew suffered little deterioration and regained preflight physical condition shortly after their return. But in the craft two of the three fuel cells had failed and the spacecraft had been operating on power from its last source when the mission concluded.

Dr. Chuck Berry, the NASA chief flight surgeon until 1974, has a picture hanging in his Houston home that captures the significance of Gemini 7. It shows a bearded Borman and Lovell walking on a red carpet from their spacecraft on the deck of the recovery carrier. The inscription by Jim and Frank reads, *"The day the straw man fell down."*

From the beginning of our astro careers we were deluged with fan mail from the curious, the well-meaning, and the career autograph collector. One of the first things we did after reporting to NASA was to sit for an official photograph. They printed tens of thousands, and the mailroom staff was kept busy filling requests for them.   •

Our egos were constantly being massaged, fed, pampered, and pumped up. This was especially true after being nominated for a space flight. *You were going up.* You were discovered. It was as though you had just been chosen the next Miss America.

In the phase that preceded this we were little more than a number. Alan Shepard, Scott Carpenter, Wally Schirra, all the Mercury types would be receiving invitations to various functions because they were Shepard, Carpenter, and Schirra. The rest of us might as well have been invited by our gym locker numbers (mine was 24). It was sometimes awkward as hell going to some high-powered society function to be received as celebrities except that no one knew us.

We could be shaking hands with a line of people and I would say, "Hi, I'm Walt Cunningham, this is my wife, Lo." But the listeners had already turned off their memory devices. We had been logged as "Astronauts," and by comparison a name didn't matter. Behind us came another astronaut, and behind him another, and another, like so many skeet at a gun club. We were looked at as a title and an image. It was a strange and sometimes unsettling feeling.

From the moment the headlines appeared in the Santa Monica papers—"RAND Scientist Selected As Astronaut"—I was aware of the impact the news was having. It was amazing how many old friends wished to renew acquaintance, how many people became old friends, and how many strangers offered free advice.

I also received my first crank letter, a paste-up of words cut out of a magazine. It carried a hostile reference to my going into space and working for a war machine like RAND. I turned it over to the security people and thought no more of it.

A small percentage of the mail came from the kind of kook who is constantly solving the problems of the universe. I got one letter that went on and on, for twelve pages, dealing with the discovery of a single unit that unified all fields of physics. I never cease to be amazed that some guy in blissful ignorance of the laws of physics can sit in his home or office and dream up a tiny unit which is supposed to open up all the secrets of mankind.

Overnight we all became collectible items. Soon after Lo and I settled in Texas we were invited to a fund-raising reception for Planned Parenthood in Dallas. It was at the ranch of Gordon McLendon, a flamboyant Texas millionaire who owned several radio stations and had run unsuccessfully for governor. There were several of us token astronauts (the old hands had all ducked) at the party, which also included several Hollywood types to help the fund raising. John Barrymore, Jr., was there. So was George Hamilton, whose romance with Lynda Bird Johnson was not yet making the gossip pages. Actor Robert Cummings was the guest contributor of a short plane ride around the ranch, all in the interest of sweet charity. Lo, so help me, was thrilled to death at the prospect of flying with Robert Cummings as the pilot.

I guess that was fair. After all, her husband and his friends got their kicks turning the Los Angeles–Houston airway into a jet age raceway.

It was inevitable that any time two or more astros engaged in the same activity a contest would result. We would compete to see which plane got airborne first, who refueled the fastest, and who arrived first at the ultimate destination.

Once Wally and I were headed for Los Angeles and I was flying the first leg to El Paso. As we taxied in for refueling, Jim McDivitt and Rusty Schweickart taxied out, taking off for Los Angeles, and we waved. As juvenile as it may seem, that's all that was required to start a race. We refueled and, with Wally now in the front seat, took off in hot pursuit, seven minutes behind Jim and Rusty. We were off and running on a 700-mile leg, hell-bent to make up those seven minutes and beat them into the parking ramp at Los Angeles International Airport. With less than ninety minutes to work with and pushing the plane as we'd have to, it was a cinch there'd be little fuel to spare.

Well, Wally gave it the old wild-blue try. It was a two-burner climb to 43,000 feet and then a level cruise at Mach .98 (98 percent of the speed of sound). Over the radio frequencies used by the Flight Service Stations, we could hear Jim and Rusty calling in along the route, as they could hear us. We tracked our progress with respect to their position reports. Hoping not to provoke them into any special efforts, we did our best to lead them to believe that we were not gaining on them. Keep in mind that no one had explicitly issued a challenge or even mentioned the prospect of any kind of a race to Los Angeles.

But that didn't matter. We just assumed that Jim and Rusty were doing their best to mislead us into thinking they were making progress no faster than usual.

Outside of Phoenix we played our hole card, going into afterburner and boosting our speed to Mach 1.2. We stayed supersonic for the next twelve minutes or as long as we dared and still have enough fuel for an idle descent down the instrument landing glide slope (ILS) and the landing—if we had calculated correctly.

By now we realized that even our supersonic fuel-gulping tactic and a high-speed letdown wouldn't enable us to land ahead of them. Our attention turned to landing as short as possible on the runway, making the first turnoff, and possibly beating them to the parking spot. For all our efforts, we had gained only two minutes in the flight from El Paso and touched down five minutes behind them, beaten cleanly.

We got high on such encounters. That undeclared contest occupied both crews throughout the flight and it dominated the conversation between the four of us all the way to the hotel that night.

The lengths to which we went to beat each other were epitomized in what came to be called the Tire Biter's Award.

To begin with, one must understand that trying to make an early turnoff on a landing runway—as a means of beating your buddy to the parking ramp—was no small matter. It was a comic-serious business to win at anything, even if it was just parking the airplane.

The T-38 could be stopped in a very short distance, but it required a heavy and precise foot on both brakes. The most competitive were always willing to risk a blown tire for the pure joy of beating the other guy. This led to many blown tires and the creation of the special award for excellence in the art of tire biting.

The Tire Biter's Award was born as a result of a little gamesmanship at Los Angeles International. Dick Truly and Bob Crippen landed one day just behind Joe Kerwin and me, and tried desperately to beat us to the chocks. After landing Joe and I rolled to the end of the runway for a normal turnoff before we realized that Dick and Bob were trying to beat us to the parking ramp. As I turned to head back for the parking spot, really balling along the taxiway, I noticed that Dick had stopped back toward the landing end of the runway. His problem became clear when we heard the call that is common around a military airfield: "Say, tower, could you have them send out a tow vehicle and a tow bar, I'm shutting down on this taxiway. I 'think' I have blown a tire."

Sure enough, Dick had made the very first turnoff at Los Angeles International, which is no mean trick. In order to accomplish it, he had worn down both main tires almost to the rim, naturally blowing them in the process. It was fortunate he was able to turn the airplane off the runway, it would have caused havoc with the commercial traffic landing behind him.

I didn't tell Dick I had quietly obtained a promise from a ground crewman to save the worst of the two tires. Believe me, they both looked like something you'd step over at a garbage dump.

We gleefully took our trophy back to Houston, mounted it on a plaque, designated it the Tire Biter's Award, and voted unanimously to make Mr. Richard Truly the first recipient. At the next pilots' meeting we had a formal presentation where Dick sheepishly accepted that creative example of one of the prices of competition. Over the years, the award would pass through the hands of many deserving tire biters.

Of course, it made no impression that we were risking our necks as well as wasting the taxpayers' dollars in pursuit of our own muse. This was how we relaxed, let off steam, and exorcised our demons.

El Paso was the scene of yet another of our private test matches. That one was for the fastest refueling stop and turnaround, as measured by time spent on the ground at El Paso International Airport.

The transient service crew there was almost all Chicano, and we would usually chat in Spanish. They had a great esprit de corps and enjoyed our flying stops as much as we did. We were always given a preferred parking spot, directly in front of the Operations Building, and more often than not they had a refueling truck standing by as we pulled in.

The boys in El Paso prided themselves on being the best in the country at refueling transient aircraft. Their efficiency encouraged us to think in terms of new Olympic refueling records. I was no stranger to El Paso, having refueled there since my first cross-country flight in 1954.

A normal T-38 refueling stop went something like this: park in front of the building where the refueling truck was waiting, come leaping out of the aircraft while they were connecting the hose, sign the credit slip, drop a dime in the coffee machine, duck into the john to deposit the last cup of coffee, pick up your fuel receipt, try to swallow a half cup of coffee, take a quick check around the aircraft, and climb in as the fuel truck pulls away. By that time they had connected a starter and we could crank up and taxi out. Total elapsed time: ten minues.

When this particular competition was in full sway, corners were cut wherever we could. Gus Grissom and Frank Borman, among others, achieved notoriety for "hot refueling": attaching the hose and refueling the aircraft while the engines were still running. While this is standard practice aboard aircraft carriers and on certain navy bases where they are prepared for it, it is clearly frowned upon at all other aircraft establishments. To this day, I think Gus Grissom holds the record for an El Paso turnaround using this technique: a flat five minutes.

The race that best reflected the true competitive instinct of the astronauts, however, didn't match us against each other. It was the nonstop flight between Los Angeles and Houston. That won't seem a very stimulating contest unless you understand that success depends not only on your own skill, but also on many factors over which you have no control, such as the weather at the destination and the tailwinds you can pick up along the way. Failure, needless to say, could ruin your whole day.

You can find a lot of air force jocks who will argue that a T-38 can't fly 1,380 miles without refueling or remain airborne for more than two and a half hours. We demonstrated otherwise, time and again, in a contest that matched man against machine and man against the elements.

We played the game in late fall and early winter, when the jetstream had dipped far enough south to give us the kind of help we needed along the route. Of course, that time of year also marked the start of the night ground fog season at Houston's Ellington Air Force Base, an added hazard to sweeten the test.

For this trip, the important part was *not* seeing how fast you could depart from the Los Angeles airport, but minimizing the time spent on the ground and the fuel used on take-off. Since the distance was beyond that which could be filed legally under the federal air regulations, a flight plan would be filed for El Paso. Fuel and ground speed checks were a continuous activity until passing Tucson, Arizona. At that point, if the necessary tailwinds had developed and if the weather was good enough for a visual descent, it was possible to legally (almost) re-file for Ellington Air Force Base, our home field.

The fuel-saving techniques we practiced were those that any good aviator acquires over a period of years. If it still seemed prudent to hedge the bet after passing Tucson, you could refile for Austin, allowing the final decision to be postponed for one more hour. The game had a way of fully testing one's nerve and judgment because landing under your own power depended on your ability to project fuel, winds, and weather conditions one hour and 700 miles farther east.

I don't recall who started this madness by making the first nonstop run, but I suspect that it was John Young or Gus Grissom. I do recall that Gus and Roger Chaffee came closer than anyone else to not making it. One of their engines flamed out on landing, and the tanks required more fuel than the book said a T-38 would hold. Deke Slayton and his NASA flight operations people raised unshirted hell over that episode. But far from discouraging the rest of us, it merely stimulated our glands.

In seven years of flying the T-38, I have made this trip more than two dozen times, and it has rarely been a boring flight. Usually it was quite the opposite, a continuing series of go–no-go decisions. One trip that stands out was a flight with Wally on a dark summer night; Wally was at the controls and I was in the back doing the navigation and fuel calculations. We operated on a rule of thumb that if we had an average tailwind of 70 mph for the duration of the flight, then we could make it to Houston—assuming, of course, we hadn't wasted fuel on take-off and climb-out.

On that night as we crossed the Colorado River heading east at 41,000 feet, we had about 30 mph of tailwind with an increase predicted as we moved toward Texas. We were pulling down 40 mph as we passed Phoenix, Arizona. By Tucson the tailwind had picked up to 50 mph. As we passed El Paso, our ground speed was 670 mph, indicating that we were now getting our 70 mph of tailwind. We knew we had to pick up at least 90 the rest of the way in order to make it home with enough fuel for an idle descent and landing.

Well, faint heart ne'er won fair lady—nor a race against the clock. I refiled for Houston and we discussed the fuel situation for the next forty-five minutes. The first low fuel warning light came on while we were still 250 miles west of Houston, indicating we had twenty to twenty-five minutes' worth of fuel remaining. I made one more calculation, told Wally we could make it if we had no problems, and put away the calculator.

We started our letdown 100 miles out with both engines in idle and a few minutes later agreed that we'd be better off to shut down one of the engines. Nearing the field on a straight-in, idle approach, Wally, not wanting to get too low too soon, was still 5,000 feet in the air. He had to make an overhead circle to lose the altitude.

The tower operators at Ellington were aware of our little gambits and generally had a good idea that we were low on fuel.* We started up the second engine while Wally was banking around for the approach. With both engines in idle, Wally was still high and diving for the ground in an attempt to avoid another go-around, a somewhat precarious situation in a T-38.

At this point I was sitting in the back seat telling myself that if he had to take it around again we just might have enough fuel for a very tight 360. At the same time, though, I sat there with my back rigid, my head against the headrest, and both hands gripped firmly around the ejection handles—the prescribed posture for an ejection.

When it came to stick-and-rudder work, Wally was one of the best. He made a beautiful landing. As we taxied in, patting ourselves on the back for making it in one piece, I calculated our remaining fuel to be less than five minutes.

Those were the hours and the personal contests we liked best. We knew NASA frowned on it and did their best to stop such mischief. But hanging our fannies over the gaping precipice and pulling them back in safely reinforced that feeling of invincibility residing in any good fighter pilot. It further increased self-confidence and, in doing so, added to that arrogance which is born of confidence. Climbing out of the cockpit after a trip like that, we felt we had squeezed something out of the machine the engineers hadn't meant to ask of it. A contest had been won not only over the machine but also over ourselves, because the greatest obstacles to performance are those hurdles we build in our minds.

One problem with our T-38's is that there were no provisions for

---

* Which was a good thing for us. It would be a hollow victory if we had to declare an emergency to obtain priority for landing.

baggage of any type—for one let alone two of us, which was often the case.

It wasn't unusual for two astros to be on the road for a week or more, complete with luggage and a full load of paperwork. As soon as the T-38's began arriving at NASA, we had to devise ingenious solutions to the cargo problem. The first was to empty the parachute seat pack—containing such survival trivia as radio, food, water, and medical dressings—and replace it with such essentials as shorts, shirts, and shaving kit. The survival gear we simply left at home. Al Shepard had caused a minor stir when he walked out to his airplane one day carrying a new attaché case that almost perfectly filled the seat pack. It was practically custom-made. Al only had to stand on one corner of the seat pack to close it. That was class. It took me a month to track down the source and acquire a similar case.

Since most astros traveled with some form of small suitcase—preferably soft—a suit bag, and a briefcase for paperwork, other places clearly had to be found to stow the gear. The suit bags were generally rolled up and placed above the ejection seat just behind the pilot's head in either cockpit. It was adequate if you didn't mind only three feet forward visibility from the rear cockpit and you liked the slept-in look.

If you really wanted unique custom luggage, you used the seat pack itself as a suitcase. Since it was designed to carry survival equipment, and to take the shock of an ejection, it attracted many an interested stare when an astro appeared in a hotel lobby in a blue flight suit and a seat-pack suitcase at his side. And, since it also served as the ejection seat pad, taking it with you was a good way to be sure no one would go flying off in your airplane before you were ready to head home.

Eventually, NASA bought a half-dozen center-line gas tanks which had been modified to serve as outside luggage carriers. It signaled an end to wrinkled clothes but also cost us ten minutes of fuel duration, and we were never completely confident of them. The tanks had been known to rip away at odd times, and we didn't relish arriving at our destinations without clothes and shaving kit. In the cockpit, at least, we were secure in the knowlege that if our wardrobe was lost, we were too. We did occasionally have visions of having to ditch our T-38, and a search party finding the cockpit absolutely awash with shirts, underwear, shaving cream, and reams and reams of paperwork.

In the face of all this—the social ramble, the home tensions, the instant hero worship—we never lost sight of one thing: *the mission came first*. Where it counted, deep in our guts, nothing else existed. Everything else was parsley.

As the months and seasons of 1966 flew by, an ugly realization was sinking in. The spacecraft, as far as we could judge, left a lot to be desired. As the launch date for Apollo 1 drew nearer, we got into one wrangle after another.

The engineers at North American Rockwell had never before been in the position of dealing directly with the user. Now here we were, messing not only with their systems, which were being tested for the first time, but also with their designs.

And there was a continuing process of redesign going on, and parts being retrofitted. If a test revealed a weakness that couldn't be fixed before the spacecraft was to be shipped, the equivalent of a repair kit would be sent along later to the Cape to make the necessary replacement. Staring us in the face was the fact that we were never going to have a complete spacecraft with all the right parts operating at one time until maybe right before lift-off.

Scraping nerves even rawer were the Apollo crewmen who had been involved with Gemini. They kept comparing North American Rockwell's efforts unfavorably to the equipment, controls, and procedures in vogue at the McDonnell plant. And before long every office at the Cape was getting into it. Dr. Joe Shea's program office, Chris Kraft's flight controller shop, our guys from the flight crews, the contractors and designers—all went through a crash series of meetings and conferences. The astronauts carried the least real authority but were the most vocal in their reservations. We argued for more safety in procedures and a better margin of error.

The changes we kept arguing for often wound up on the desk of Joe Shea who, as program manager, had to make a decision on how much to bend the budget and how far to push the schedules. It was natural that on many of our exchanges we felt that Joe was on the other side of the issue from us—lined up with the contractors.

Joe was a bright young man, a contemporary of the astronauts, and he affected many of their habits, including the casual Ban-Lon sportswear and their competitive attitudes. He was an incurable punster whose raised trouser legs—when he rested his feet on his desk—always exposed red socks. On a handball or squash court he would scrap with the best we had: Mike Collins or Rusty Schweickart. He was tough and active and ran his operation as jealously as we expected to run our spacecraft.

We respected Joe and liked him, and it was ironic and sad that our efforts so rarely received his endorsement. Where many of the other program managers caved in when the astronauts insisted on a change, Joe held firm. We failed to get through many crucial changes simply because we couldn't sell Joe when the cost or (especially) the schedule were affected. Somewhere along the way we began to accept the fact that the spacecraft was less than perfect. What the hell, we'd make it work anyway. Anything to get off the ground.

Through all this, one thought continued to trouble me. My Apollo 2 mission was clearly redundant, Gris Grissom and his crew would prove that the spacecraft could fly and support man in space. If Gus accomplished what was expected, our flight would be as useless as windshield wipers on a submarine. The only thing Apollo 2 had going for it was the fact that Gus kept throwing the scientific experiments off Apollo 1 and they would be rescheduled on our flight. Gus really wanted to make a pure engineering test flight and Apollo 2 was picking up what both Gus and Wally considered "junk," which translated into "anything scientific."

Wally, for the first time in his life, was faced with flying a mission overrun with science and "dumb shmuck" scientists. We had a scientific airlock on board, the same one which would be used seven years later on Skylab. And there were biological tests, medical experiments, and all the laboratory jazz that over the years had bored Wally. Now he was inheriting it all and taking it with a smile. I didn't learn why until a year later, when Deke confirmed he had planned for Apollo 2 to be his own mission. All he needed was medical clearance.

It was the one way our redundant mission made any sense. We figured to establish little that couldn't be learned from Grissom's flight. That made it low profile with little risk to the program if it didn't succeed. It was an ideal spot for a pilot trying to get ungrounded from a questionable heart condition.

Slayton was the sentimental favorite of the original seven astronauts and the only one never to have ridden a rocket. Two months before he would have become America's second man in orbit, his world collapsed. He was grounded! It was a setback which, at the time, didn't seem permanent. Deke fought that judgment from the beginning and was still fighting it in January 1966 when Schirra was named commander of a flight in which he never seemed to have his heart. The explanation was simple, he didn't expect to fulfill it. He was to be the caretaker commander, standing in only until Deke was approved for the mission.

I have often wondered how different the careers of Eisele and Cunningham would have been if Deke had pulled it off. When he got the final turndown in the early fall of 1966 the reaction was all Deke. He headed for Alaska and a couple of weeks of hunting.

Wally, who had never been up longer than a day, was now stuck with a mission scheduled for eleven days in orbit. And he was nursemaiding two rookies who could care less how unimportant the flight might be. We just wanted to get off the ground.

On top of this the hardware was bad, the testing was slow, the procedures were new and incomplete, the crew training seemed woefully behind schedule, and it didn't seem like we could get there from here. With all the problems and work we were wrapped up in, we needed to spend some valuable time away from the office and telephones to hold one of Wally's famous seances—actually a conference or retreat—and discuss some very important matters. With all of our training obligations it was virtually impossible to get the nine members of our prime, backup, and support crews all at the same place at the same time, unless it was specifically scheduled. We hadn't all gotten together away from our offices, contractors, etc., for six months, and we were overdue. The idea was not only to have a place where we could get some quiet concentration during the day, but one where we could relax, lift our glasses at night, and have a good time together as a crew.

Mid-September 1966 found all of us flying into Miami for two days of work and play. A friend offered us the use of a house with a swimming pool. It was ideal. We could work around the pool in the daytime, and at night we were right across the street from some swinging spots. This was the kind of "even strain" that Wally believed in. All nine of us in a fun city working and playing together.

We had a good time at night all right, but whatever the original intent of our daytime sessions, it got lost in what was produced at the end: a two-page list of ultimatums to the Apollo program office. Many were legitimate concerns already shared by the program office; some were "motherhood"; all were ultimatums ("If this doesn't happen—we won't fly"). Joe Shea wasn't the kind of man to listen to ultimatums, especially from a flight crew with a vested interest. We had some neat little thoughts like:

1. Delete the planned rendezvous exercise,
2. Delete certain major scientific experiments,
3. No more assigned extra duties,
   and thirteen other items of lesser significance.

We titled it a "crew list," but it strongly reflected Wally's personal preference in the approach to problems. Donn and I were concerned and nervous as hell about the possible repercussions from the moment it was drafted. We knew we had overstepped the bounds of propriety and proper procedure and were waiting for the other shoe to drop. (Which it did.)

Wally saw it as a test of who was in charge. He was prepared to defend every single item on the list to a far greater degree than he actually believed in it.

Weeks passed and we heard no direct reaction to our Miami manifesto—and no action either. I had begun to think that we were in smooth waters.

On November 11, Jim Lovell and Buzz Aldrin lifted off in Gemini 12. Four days later Wally, Donn, and I sat in the third row of the theater seats at Mission Control watching the final splashdown of the Gemini program. Sitting directly in front of us were Paul Haney, the public affairs officer, and some of the people from the Program Office at NASA headquarters. Paul half-turned toward Wally and handed him a memo.

"Say, Wally," he said. "Do you want to put your initials on this before we release it?" Paul explained later that he had wanted to cut us in before we read it in the paper.

I watched Wally as he read it quickly. His face flamed. The memo was a press release explaining how the Apollo flights were to be rearranged. It included the news that the second manned Apollo mission was canceled. Even though Wally hadn't exactly been thrilled over making the flight, he wasn't prepared to have it taken away from him, and certainly not in this manner—being asked, after the fact, to initial a press release.

Wally motioned Donn and me out to the hall. Wordlessly, he handed us the release. He was furious. We were stunned, speechless. I couldn't help but wonder if Wally's ultimatums—even though some of them were justified—hadn't painted us into that corner.

That night, in the only instance I can remember in all the time we were together as a crew, we went across the street to Nassau Bay Inn and drowned our sorrows together. If we didn't get completely pie-eyed, it wasn't for lack of trying.

In the shake-up the McDivitt crew was assigned to the second Apollo flight, which would now include a Lunar Module. It would also be the first Apollo rendezvous, and Wally could really have gone for that. He had his sights set on it, in fact, when we prepared our ultimatum. What happened instead was the first break in the pecking order between the Mercury and Gemini astronauts.

In a few days word came down that we were to replace McDivitt's bunch as the backup for Apollo 1. There was logic to this since we had lived with the only other spacecraft identical to Grissom's. There was one hangup. As a Mercury astronaut, Wally felt he should not be put in the position of backing up anyone ever again. He was adamant about it. Donn and I were just glad to still be in the flight rotation.

So Gus, and NASA, had a problem. It was now mid-November, and Apollo 1 was to fly in February. A backup crew was essential. An experienced backup crew was desirable. An appeal, a very personal appeal, from both Gus and Deke was made to Wally. I mean, they damn near got down on their knees and begged him. Reluctantly he gave in.

It was clearly a letdown for us. We were faced with three months in a backup role, and then what? We were in limbo. I alternated between feeling betrayed, by fate and by Wally, and feeling angry.

He probably couldn't have prevented the cancellation. But we would have preferred not to have him pounding nails in the coffin.

Astropolitics

NOT GIVEN LIGHTLY to introspection, we tended to accept the judgment of the press that astronauts were "mission-oriented." That mission: to be up front when the line began to form, to haul ourselves into space and get there before the next fellow. Oh, we were that all right.

But we really operated on two separate emotional levels. Sure, we were swept up in the spirit of the moon race, but back at the office, we were grubbing around in the pits, doing our damndest to keep everyone else's footprints off the backs of our necks. Each of us was learning how to play the space game. It was brand-new and the rules were made up as we went along.

When a man was selected as an astronaut people thought of him, if they thought at all, as having arrived at the very crest of his chosen profession. In one respect, he had. But at the same time he found himself standing on the first rung of an entirely new profession. All of a sudden he was back at the starting blocks, surrounded by people who had been screened, selected, almost bred for their motivation and competitive spirit.

Keep in mind that NASA's astros had come from an environment where, by natural selection, they had been or were bent on becoming leaders in their activities. Whatever the crowd, they had stood out. Their performance had made recognition commonplace. Now they were in the space league, Astronaut Division, where there was only room for two or three chiefs. All the rest had to be Indians.

Combine this with the fact that the stakes were high—glory, respect, admiration, fame and fortune, and even a place in the history books— and the place could, and did, turn into a pressure cooker.

As each man measured his career he understood and accepted these conditions. We weren't naïve, because none of us had time to be naïve. We all recognized the significance of making the first lunar landing long before the Apollo program got off the ground. If it wasn't your first conscious thought upon being selected, it hit you soon after reporting to the Astronaut Office. It was the boldest adventure and the biggest plum since Columbus, and we all knew it.

If someone takes a gang like this—screened and selected from the cream of America's manhood—and sits on them, smooths them, dewrinkles and presses them, attempts to make them indistinguishable from each other, what happens? They accept the obvious challenge and do what comes naturally: attempt to stand out from those around them.

Crew assignments were all the incentive we needed. Once they came within our reach it was like waving a piece of raw meat in front of a cage of hungry lions.

Jockeying for position became a constant activity. The game was to move ahead or—just as effective and often easier to accomplish—move the other guy back. It was a competition guaranteed to bring out the worst in a guy. I know of no astro who, upon leaving the program, didn't breathe a sigh of relief at dealing himself out of that part of the game. They may have loved the job, the competition, and some of the men with whom they competed, but the dog-eat-dog atmosphere that was part of the daily existence caused even the best of gamesmen to be glad when they were rid of it.

Some of the rules were similar to office politics in any corporate structure. But others were unique to the life, times, and men in the astronaut program.

Wherever two or more ambitious people collect, competitive situations develop. That's business. That's sports. That's politics. That's life. In our situation it was compounded not only by the military backgrounds, but also by the complicated caste system that had come into being.

Our opportunities were influenced by seniority, which in the Astronaut Office was most akin to military rank, and by the pecking order. When I joined the program there were twenty-seven people playing the game, not including Deke Slayton, who as director of flight crew operations floated above the sea of politics, or Al Shepard, whose inner ear problem kept him grounded, or John Glenn, soon to retire. Of the twenty-seven, only four were civilians. And of those four there were two—Elliott See and Neil Armstrong—whose civil service rank actually surpassed the highest equivalent rank of any of the military astros. At the other end of the spectrum were the other two civilians, Rusty Schweickart and Walter Cunningham. In between were strung out all the military people, with their own brand of pedigree: rank, date of rank, and a lineal list in which they also compared themselves according to when they graduated from West Point or Annapolis.

As the senior military man in our third group, Dick Gordon was automatically singled out by Deke as our spokesman. In fact, in terms of military seniority, Gordon had numbers on some of the astros selected earlier. But that ran headlong against the most important astronaut fact of life: the pecking order was more important than seniority.

The pecking order came about from several factors but was primarily related to the hierarchy of selection. If one were a Mercury astronaut presumably only God and James Webb held higher positions—and that might be subject to argument. Then came the Gemini group, whose members outranked and overruled all the weenies and plebes underneath them; meaning us, the Apollo astronauts. That pecking order prevailed throughout the system. In fact, it *was* the system. The only official recognition of this system came in an April 1964 memo from Slayton classifying the thirty of us as command astronauts, senior astronauts, and astronauts.

Later there would be de facto subdivisions to cover those who had been in space and those who had not; those who had been to the moon and those who had not; and finally, those who had walked on the lunar surface.

In the beginning, though, it had been a clean and simple world. There were seven, "we seven," all-for-one-and-one-for-all. They paraded together, went to the White House together, and lunched with the Shah of Iran together. They were a package.

Then the ranks grew. There were sixteen, and it became necessary to find a way of dividing the good deals, and even the bad ones. The addition of the Gemini boys created a logistics problem. There wasn't always room for sixteen. The whole crowd could no longer go to the Kennedy compound.

Within the Gemini group, leadership positions were quickly assumed by Frank Borman, Tom Stafford, and Jim McDivitt. That was no doubt partly on merit and partly on the first impressions of Deke and the original astros. As time passed, it grew more and more evident that first impressions were almost impossible to live down. Deke's first assessment of each man—even as early as the selection process—dictated what happened to him for a long time to come.

By the time the Apollo gang of fourteen was initiated, the system was fairly pat. It wasn't long after we arrived that Gus Grissom and John Young flew the first Gemini mission. John, on his return to earth, publicly announced what we had already come to believe: "You're not an astronaut until you've flown in space."

It worked this way: If an astronaut had been in space, he was a star. If he was on a crew, he was a prospect. If he was not yet in line, he was simply a suspect. He hadn't really made the team.

In the case of Rusty and myself, we weren't even called astronauts when we checked in. For our civil service records we were given the job title of "aerospace engineer and pilot." But rest assured we never filled out a form or signed a letter or completed a questionnaire without writing *astronaut* in the space reserved for "occupation."

In those early months we were divided into four flights, each with one of the active Mercury astros as chief: Gus Grissom, Wally Schirra, Scott Carpenter, and Gordon Cooper. Not only did they have the name and the game, they had considerable political clout on which to a large extent the Astronaut Office depended to get its way.

As time passed I could see certain forces at work that would change the order of things. For one, we felt Scott Carpenter was being eased out. He still had his place in the pecking order, but had inspired some disaffection among his fellows and, more to the point, among the upper levels at NASA. I had been surprised at the sniping that went on when John Glenn withdrew. Now I saw the isolation that began to surround Scott Carpenter, triggered by the fact that we, his peers, felt he had done a less than professional job in his first—and only—flight.

Scott's was an unusual case, the one Mercury astro who really marched to a different drummer. He was independent, esthetically oriented, with a bit of the poet in him. On his flight in Aurora 7, Scott had become absorbed in watching the beauty below and photographing the "firefly" particles that flew off the spacecraft. He was on a kind of natural high—and he didn't maintain the proper attitude for firing his retro rockets prior to reentry. He missed the splashdown zone by 250 miles and, for a few suspenseful moments, the very real fear existed that he would be America's first space fatality.

Gordon Cooper, at the tracking station in Guaymas, Mexico, followed Scott on a repeating set of instruments. As Scott came over the horizon from Hawaii on his last pass he was apparently well behind the retro time line. Gordo watched his friend fire the retro rockets late and in a bad attitude. Scott was using up his reentry fuel rapidly, holding his attitude badly, and swinging so erratically no one could predict where he'd land.

At one point Gordo pushed his chair back from the console. Tears came to his eyes and he buried his head in his hands. He was certain his pal was going to be burned alive on reentry.

But Scott made it down, and an hour later the first aircraft on the scene found him blissfully at ease in his raft, enjoying the sea breeze, his mind years away. He was a lot less concerned about himself than Gordon Cooper had been a few hours before watching him wobble over the horizon.

None of this was bandied about outside the family, and not a great deal of it on the inside. It was part of the code that astros did not ask each other point-blank about professional screw-ups. Scott received the usual medals and public back-pats, but within the group it was gradually understood that he had made his last flight. The word got around that Chris Kraft had told him flat out, "You're never going to fly for me again."

What followed for Scott Carpenter had to be a living hell. It had taken him fifteen years in preparation to reach that moment on May 24, 1962 when he was literally on top of the world. Then it was gone. There was nothing left.

Scott was a fine, attractive fellow, one of the easiest of the Mercury bunch to get to know. He was a little too much like John Glenn for some of his contemporaries to feel comfortable with, but always an interesting character. He looked like the athlete he was. He had been a gymnast and a track man, had good reactions, and was in the best physical shape by far of any of the Mercury guys. But he wouldn't compete. Where the others took up handball, Scott would go for something more esoteric. When we opened our new gym, he took up fencing and invited his instructor out to join him. I remember seeing them there, uncasing their foils and putting on their wire masks. He looked like Captain Blood. Watching him through the years, letting his hair grow, getting involved in deep-sea diving or soaring or music, Scott always did his own thing. I respected him for it, but it didn't endear him to the group.

So Scott began to wander afield, taking off for Bermuda, breaking an arm on a motorcycle, becoming an aquanaut and manning sealab, one thing after another. One ironic fact that binds us all together was first visible in the case of Scott Carpenter: no matter how well or badly it turns out, no matter how easy or painful the going, no astro wants to say good-bye to the job. Perhaps unconsciously, Scott was expressing an attitude that persists today, in which many cling to the title of Astronaut.

I suppose Gordon Cooper was the next to fall from grace. He gained the reputation as another guy who followed his own muse. He, Gus Grissom, and Jim Rathman, the former Indy winner, formed the GRC

Racing Team and entered a car for several years in the Indianapolis 500. They didn't have a winner, but they had the usual sponsors to cover part of the cost and they had a good time, a month-long party each May. The first time I ever went to the famous old Brickyard was to visit with them and help to work on their car. Al Worden, only recently arrived in the program, took a few weeks' leave and spent them in Gasoline Alley with the GRC Special. It was a lark, although I do think we could have made pretty fair mechanics.

But none of that endeared Gordo to the establishment. Nor did a later remark, when the brass ordered him out of a race he himself was scheduled to drive at Daytona Beach. He had already taken a turn at speedboat racing, and Gordo had a mild case of apoplexy when the order came down from NASA on the eve of the Daytona race. The withdrawal wasn't exactly graceful. "What do they expect us to be," he fumed, "tiddly-winks players?"

Cooper's rebellion was sincere, but he shouldn't have been too surprised at the repercussions.

While various internal factors were affecting the careers of each astronaut, another, more obvious, classification was being made by the general public: the distinction between those who had been "up" and those who had not. It surfaced principally when it came to filling requests for speeches or some of the blanket social invitations in the Houston area.

That was understandable. When an organization invited a dinner speaker they wanted someone who would attract as much attention as possible, a crowd-puller. It was a poor substitute to have a pretender to Valhalla stand up and tell the audience about jungle training in Panama or flying the mission simulators.

At the risk of reducing our own odds of flying, we looked forward to the arrival of the next group of astronauts so we could drop the "new" from our introduction. Having to exist in reflected glory for so long could explain why so many wasted little time establishing their own identities after their flights.

Of course at first we were so gung-ho just to be there, it didn't matter what we were called. I recall visiting NASA about six weeks before I had to report, hopping into an A-4D and flying from California with Hugh Purser, a friend from my marine reserve squadron.

We parked with all the NASA aircraft and made our way to the flight office where the fellow in charge assigned us a locker. We could look around and see what was in store for me. We admired the custom-made, hot-looking flying helmets—custom-molded to each astro's head. And custom-made boots. The flight suits were NASA blue, not one of those low-rent orange or green ones that military pilots wore. It was clear that I was going to dress first class. I was joining the last of the great flying clubs in the world.

Purser was dutifully impressed that they would roll out the red carpet for a weenie like me. But the topper came when we walked through the reconditioned barracks that housed our offices and I casually waved and said hello to Al Shepard and John Glenn and the rest of the legitimate heroes. It was a sweet, heady kind of atmosphere. On that visit I not only couldn't tell that a pecking order existed, I also wouldn't have given a toot that it did.

The illusion of quick acceptance was to be heightened by the genuine courtesy of John Glenn. Very soon after I joined NASA, John dropped by and asked if I was having any problem maintaining my marine reserve affiliation. I had transferred to a Dallas squadron and was looking forward to flying the FJ Fury. There was still an uncertainty as to whether I'd have the use of a NASA plane to fly from Houston on weekends and be able to keep the Marine Corps happy with my attendance.

John offered, "A good friend of mine, Jay Hubbard, is an aide to the commandant. If you have any problem with the Marine Corps, let me know. Jay will take care of you." As a matter of fact, Jay did, and over the years, through John, became a good friend.

All any of us wanted was a space flight, and we rarely pretended otherwise, at least among ourselves. I doubt that any of us, in our hearts, were any more concerned with the good we were doing mankind than Indy 500 drivers are motivated because they believe the risks they take will develop safety improvements in highway cars.

For a long time many of us believed that our performance in training would determine how we fared. We learned eventually that it wasn't enough. We kept butting our heads against the pecking order. Our Apollo group had the distinction of being the first to whom Deke made a point of *not* promising a flight to every selectee.

So the pressures mounted: to get on a flight crew, to climb out of the shadows and make your own dawn, and to appear in public without being reduced to showing the latest film of the flight of your "buddy," Dick Daring.

At the very bottom of the pile were the hyphenated astronauts, the scientists, who the pros simply assumed would be unable to cut the mustard "with the aviator fellows." That rankled. But by the time the second batch of Ph.D.'s arrived a sub-class distinction was quickly established between the first group of scientist-astros and the second.

To understand how the pecking order prevailed, it's helpful to examine the careers of some who persevered. You can begin with Al Shepard, the first American in space. When Shepard's inner ear prob-

lem grounded him, there was only one logical position for him and that
was in charge of *all* the other astronauts, save Deke Slayton, for whom
the precedent had been established.

Al was placed in charge of the Astronaut Office, a division of the
Flight Crew Operations Directorate (formerly handled by one of
Deke's assistants). Al stayed there until the ear problem was corrected.
Then he successfully moved himself directly onto a flight crew. At that
point he was succeeded by the resourceful Tom Stafford, a second
generation, Gemini astronaut.

When Shepard returned from his big comeback, Apollo 14, he had
made it to the apex of the pyramid: the only Mercury astronaut to
walk on the moon. He was the king of the pecking order and was soon
to be promoted to the rank of admiral. We all knew he wouldn't wind
up working under Tom Stafford. Al took back his old desk, leaving the
question, What to do with Stafford, without making it appear he had
been demoted? Easy. He was moved upstairs as a staff assistant to
Deke.

Even the bittersweet case of Gordon Cooper was consistent with the
rule of the pecking order. Gordo flew the longest (thirty-four hours)
and probably the best of the Mercury missions. Later he commanded
the third Gemini flight with Pete Conrad as his crewmate, and set a
new space endurance record of 190 hours, 55 minutes. He was far and
away the marathon champ of all the world's spacemen.

As the youngest of the magnificent seven, Cooper seemed to have
the highest hopes of making the first lunar landing. Public sentiment
certainly favored one of the original Mercury heroes taking that historic first step. But along the way it went wrong for Gordo.

He was outspoken. He complained—in print—about some of the
NASA pressures. His reputation as a daredevil haunted him. He made
waves.

Cooper never flew in space again. His career was one long series of
being leaned on for his transgressions, real or imagined. He paid the
price for not toeing the mark.

Here was the irony of it all. Even as Cooper resisted, and the brass
above turned the screws, the pecking order was religiously maintained.
Gordo backed up the last Gemini flight and then, with Donn Eisele
and Ed Mitchell on his crew, backed up Apollo 10.

That should have put him in line to fly Apollo 13. But when it came
time to announce the crew, Al Shepard stepped forward with a clean
bill of health from the medics. He was ungrounded and wanted the
flight. Deke obliged, only to see headquarters turn it down. The headquarters feeling, shared by his fellow astros, was that Shepard needed
more time in training.

Some fast shuffling was necessary and Jim Lovell, who had been scheduled to fly on 14, was given command of Apollo 13.

Shepard went to work with a passion, and with the public rooting for him, won his case and convinced the powers at NASA to give him Apollo 14. He picked up Ed Mitchell from Cooper's backup crew and Stu Roosa, both fifth-generation astronauts.

That cooked it for Gordo. He had done some fairly colorful, behind-the-scenes bitching to his old pal Deke when he was asked for the second time running to take a backup crew on Apollo 10. There had always been that magic lantern shining in the distance, the chance to make a lunar landing. Now there was nothing. Cooper had been passed over again, and he got the message. Everyone did.

But even then, they had to find him a place appropriate to his position in the pecking order. And they did. Gordo was moved upstairs to work with Deke, whose office was a sort of halfway house for those who couldn't, or wouldn't, work for someone beneath them in the pecking order. He marked time there until July 1970, when he left the program. Although he has never said so in public, it was no secret among the rest of us that Gordo felt his moon flight had been pre-empted by Shepard.

A sense of timing was crucial to anyone who wanted to get ahead—meaning all of us. The only way to beat the PO was with the right-place-at-the-right-time stroke of luck, such as being on a backup crew when one of the prime members bought the farm, or maybe broke a leg. It was all fate. Take for example Jim Lovell (called Shakey by his friends).

When Lovell successfully completed his Gemini 7 flight, he was assigned to back up Gemini 10 with Buzz Aldrin—in what could only be considered a dead-end assignment. All the early Apollo crews would be in training by the time they became available for reassignment. The only duty worse than backing up 10 or 11 was to back up Gemini 12. (Gordon Cooper got that one.)

Enter the hand of fate. Elliott See stalled his T-38 on a routine landing and crashed into the roof of the McDonnell plant in St. Louis, Missouri, killing himself and Charlie Bassett. They were the prime crew for Gemini 9. Their backups—and also their buddies—Tom Stafford and Gene Cernan inherited the mission. That started the chain reaction. Lovell and Aldrin moved up. Lovell was now backing Gemini 9, instead of 10, which put him in line to fly Gemini 12 in November 1966.

Normally that would still have left Jim out of the early stages of the Apollo program, because of the overlap. But fate stepped in again.

Two months later, the Apollo 1 crew died in the fire on the pad, and there followed a two-year hiatus in manned space flights.

Lovell was available again. After the dust settled on the postfire reshuffling, he wound up on the backup crew for Apollo 8. The prime crew was Borman, Collins, and Anders. What more could happen? Lots! Mike Collins, originally on the Apollo 8 prime crew, developed a numbness in his arm from a pinched nerve and underwent surgery. Jim Lovell, his back up, was moved up, flew around the moon with his old partner, Frank Borman, and subsequently commanded Apollo 13, thus becoming the first man to fly four space missions.

As for Mike Collins, by the time he could get back on flight status, the next mission open was the first one bound for the surface of the moon: Apollo 11. Mike replaced Fred Haise on Armstrong's crew, and piloted the Command Module while Armstrong and Aldrin took their historic "giant leap for mankind."

Then there was Donn Eisele, who had been ticketed for a spot on Gus Grissom's Apollo 1 crew. A few days before the news was to be announced, Donn threw out his shoulder playing handball, and following surgery he had to keep his arm in a sling for several months. When the crew was introduced, Roger Chaffee had replaced Donn.

Few ever knew that Eisele had been slated for that doomed crew. I didn't know myself until months after the fire. We had worked together for a year and a half when, quite out of the blue, Donn turned to me one day and said, absently, "You know, Walt, I was supposed to have been on that crew." That was all he said. But I knew what he meant.

Timing was an element beyond our mortal control. But we were all aviators and surely, we thought, how we flew, how we performed, ranked way up there among the things that counted. But just as it was unpredictable what benefited a career, what penalized us was equally mysterious.

As aviators the astronauts had the usual number of accidents for the hours we flew, and a good majority could be attributed to pilot error: either outright poor judgment, or a bad response to a developing situation. Yet it didn't seem to handicap anyone's career. Only once—in the case of Scott Carpenter—did it appear to matter when a guy even blew a space mission.

At such times the value of personal relationships was difficult to judge. Joe Engle was an air force test pilot who had flown the X-15, an early rocket plane. Joe was a great guy with impeccable credentials, and only coincidentally one of Deke's hunting buddies. (Deke could respect you and enjoy your company if you didn't hunt, but it helped if you did. He was an outdoorsman and I suspect my stock went up a bit when he learned my half brother was an air guide in Alaska—a bush pilot who made a living hunting and fishing.)

After Joe Engle learned to fly helicopters he gained a new mobility for some of his hunting forays. Coming back alone one weekend from a hunt outside of San Antonio, Joe was running late. It was after dark when he made it back to the field—almost. A half mile short of the runway his chopper ran out of gas. He effected a successful night auto rotation—a dead-stick helicopter landing at night—which is quite a trick and one that Joe had never practiced.

The chopper went through a rail fence and all it was good for afterward was spare parts. Because Joe was a fine pilot, he walked away unscratched. If there was ever a word of reprimand, no one heard about it. Not long after, he was assigned to the backup crew for Apollo 14.

Another X-15 veteran, Neil Armstrong, parlayed a busted Gemini 8 flight into the Buck Rogers grand prize mission, the first lunar landing. It was Neil's first flight with Dave Scott and the first docking in space.

It was March 16, 1966 and my thirty-fourth birthday. Lo had accompanied me to the Cape where my job was in the blockhouse as capsule communicator for the flight crew—a position we called "Stoney," the origin of which no one knows. The primary function of Stoney was to communicate with the crew and call "lift-off," a small but active assignment that let one feel he was making progress.

It was hard not to be amused by Stoney's main responsibility: counting from ten backwards and hollering, "Lift-off." It was almost that simple. Stoney did provide visual confirmation of lift-off to the crew, *the* key piece of information in abort decisions during those hectic seconds immediately before and after lift-off.

Immediately after lift-off, as soon as I could get out of the blockhouse, Bill Anders and I and our wives hopped in a Cessna 172 and started our nine-hour, one-thousand-mile flight back to Houston. On the way we attempted to catch the news broadcasts on the progress of Gemini 8. The first rendezvous and docking in history was going to take place while we were en route home.

By the time we landed in Houston the mission was already over. Neil Armstrong and Dave Scott had been picked up by a destroyer in the Pacific and Gemini 8 became the second-shortest Gemini mission on record. It was described as the first "emergency" in space. After docking with the Agena, a runaway thruster began rotating the vehicles. Malfunction procedures had been written and practiced by the flight crews for just such an eventuality but at the onset of the problem improvisation seemed to be the rule of the day. When the excitement of the moment was over, the spacecraft was undocked and once more facing the Agena in space, but the crew had unnecessarily activated a backup control system.

The mission rules for the situation were plain: reentry at the next available recovery area. Eight hours after lift-off they sat floating in the Pacific, waiting for a destroyer to pick them up.

Of course, Neil and Dave received the usual medals. Neil went on to become the first man on the moon. And Dave Scott's career as the fair-haired boy of the third group of astronauts was unaffected. Both performed well over the remainder of their careers, but at the same time their very progress ignored the fact that their peers—and many others at the space center—felt they had botched their first mission.

In such situations, we generally felt a curious mixture of relief, pride, and envy. We regretted not having had the opportunity to demonstrate how flawlessly we would have performed under the same circumstances.

As a final example of accidents that seemed to have had no adverse effect on a fellow's career, there was Gene Cernan's mishap, which occurred at a time of great speculation on who was going to fly the last Apollo mission: Apollo 17.

Gene was the backup commander on Apollo 14, with Ron Evans and Joe Engle on his crew. Backing up Apollo 15 were Dick Gordon and his crew. We all knew at the time that Apollo 17 would be the last of the moon missions and in the normal scheme of things Gene Cernan would have rotated right into the prime crew spot. But for a variety of reasons—including having the only honest-to-God space geologist, Jack Schmitt, on his crew—some were guessing that Dick would actually get it. Apollo flights were now flying at such long intervals, it was no longer necessary to skip two missions in rotating crews.

Dave Scott, who commanded Apollo 15, even made a personal pitch to Deke to help Dick get the mission.

In the midst of this uncertain situation and one week before launch day, Gene Cernan decided to take a helicopter training flight down the Indian River, which runs behind Cocoa Beach. NASA had moved a helicopter to the Cape so that Gene and Al Shepard could keep their hands in at flying lunar landing approach profiles. For whatever reason, Gene wound up flying the helicopter into the drink.

While the accident board findings may not have reflected it, among the boys the verdict was that Gene Cernan had been flat-hatting—buzzing along the river. He may have spotted a nude sunbather, or whatever. In any case, Gene crashed into the river. The chopper turned over and only good fortune saved Gene from being killed. As it was, the shock knocked him unconscious. He came to under the water, found the bubble of his chopper broken, struggled to the surface, and swam desperately through water now wild with burning gasoline. Gene's charred helmet was found later.

Gene followed the training manual: he took a deep breath, dived, and swam under the fire, still in his boots and heavy flight suit. An old woman, fishing in a rowboat, finally pulled him out of the water.

The accident board convened, took weeks to gather its findings, took months to file a report, and finally confirmed what everyone had assumed: pilot error rather than equipment failure. The betting in the office on the Apollo 17 crew had long since switched—aviators characteristically do not wait for the accident report—"That sure cinches it for Dick," the refrain went. "Ol' Gene just screwed the pooch."

But it had absolutely no effect on the assignments for Apollo 17. Consistent with the past handling of such incidents, Gene Cernan still became the spacecraft commander for the last lunar landing.

Some might have wondered what kind of guardian angel was looking over Gene Cernan's shoulders to guide him to what can only be described as an extremely successful career—clearly the grandest of any in the third group of astronauts. The answer is Tom Stafford.

Gene Cernan was a lot of the things that Tom wasn't. Gene—already a veteran of two successful missions—was one of the few astronauts who could outdo Wally Schirra on the social circuit. He had all the essential graces. He quickly and easily cultivated the rich and the famous and, in turn, was cultivated by them. His beautiful wife Barbara was very comfortable in the same surroundings.

Tom and Gene came together on the backup crew of Gemini 9 and, after Elliott See's accident, the flight itself. It established a relationship that paid off handsomely for both of them for the next seven years.

Theirs became a friendship that can best be described as symbiotic, with Gene providing the P.R., the social entrée, and a Jet Set lifestyle and Tom providing the technical know-how, the hard work, the mission-oriented aspects of the partnership, and the career control. It was a classic relationship.

Tom Stafford came along in the second generation of astros, and from the beginning he was one of the most respected men in the program. He developed the right kind of earthy camaraderie with Shepard, Slayton, and Grissom, and rose to a position of authority over his contemporaries. This was due not only to his relationship with the Mercury guys, but also to the fact that he migrated toward positions of greater responsibility and toward the center of power. Tom wouldn't be content going from prime to backup to prime crew, ad nauseam. He had his sights aimed beyond that.

When Al Shepard finally made a successful stab for a flight, Stafford came off his Apollo 10 flight to replace him in charge of the astronaut office. The only man still in the program higher in the pecking order than Tom was Gordon Cooper. To avoid an ego crunch and maintain the pecking order, Gordo moved up to become Deke's assistant.

And Stafford became a caretaker chief, the last in a sequence of events that created at least three problems among the troops:

1. Al and Tom's good fortune was clearly Gordo's loss and even those not close to Cooper were sympathetic on that point.

2. A majority of the astros resented not only Shepard's desire but also his ability to place himself directly on a flight crew without even the pretense of embarrassment at skipping a backup slot or of regard for those who were cut out of the pattern in the process.

3. Some of the second group of astros, Tom's contemporaries, were less than thrilled at his rise to a clearcut position of authority over them.

While it was a paper-pushing job without glamor, Stafford was nicely located within the "corporate" structure, and in a unique position to dispense the space center version of patronage. He could influence which public appearances were approved around the country, who was assigned world-tour boondoggles, and even exercise an input on flight crew assignments, including his own.

Stafford kept Shepard's chair warm for nearly two years while Al made the third lunar landing. When Al returned Tom joined Deke's staff and began laying the groundwork for his own power play, à la Shepard.

The only mission projected for the future without a flight crew assigned was the Apollo-Soyuz Test Project scheduled for 1975, and Deke had his cap set on both flying again and commanding that particular mission. Well, so did Tom Stafford, and he had learned the rules and how to play the game like an expert. After all the infighting was over, the public announcement listed Tom Stafford as spacecraft commander with Deke Slayton, Tom's boss of ten years, working for him.

Stafford, the acknowledged master at playing astropolitics, is a tall, balding, slightly awkward-looking fellow with a Lyndon Johnson earthiness and a capability for the technical aspects of a mission. He was ambitious and smart, except when it came to investments. He probably took more chances on dry holes and poor insurance companies than most of us.

Tom was deceptive in the impression he made on people. He looked like a Presbyterian minister and he played the "good ol' boy" role to the hilt, notably when he was testing the political waters around the state of Oklahoma. But it was a mistake ever to underrate him. He was perhaps a better chief of the Astronaut Office than Shepard. He devoted more time to it and generally represented our interests well.

With all his attributes, a social whiz he was not. He knew how to have a good time but wasn't much for the social charade. Thus Stafford and Cernan blended their talents perfectly.

Although Cernan prevailed over Dick Gordon on the Apollo 17 mission, one member of Dick's backup crew made it: Jack Schmitt. That was one situation that was *not* resolved according to the pecking order.

In his own way, Schmitt was an interesting case. Jack was movie-handsome, dark, and short (two out of three isn't bad). He had spent a lifetime outdoors, had a good reputation as a geologist, and was considered by some to be brilliant. Jack and I had met prior to his selection as an astronaut. He was working for the U.S. Geological Survey in Flagstaff, Arizona, at a time when one of my astronaut duties had been to work with the USGS on lunar surface activities.

Jack and I didn't hit it off particularly well in our early encounters. He had quizzed me about becoming a scientist in the space program and I had not been very encouraging. It was clear that he didn't appreciate my attitude, which was that the space program wasn't yet in need of pure scientists—an opinion borne out by subsequent events. On the other hand, I sensed that Jack felt the skills of the aviator were overrated and I resented that.

When NASA selected the first five scientist-astros, Jack was right there among them. He had an uphill fight all the way and no one worked harder for his seat in a spacecraft. He finished his year of flight training with the rest of the scientists and graduated definitely without honors. But he kept at it until he became an able, if conservative, aviator. Nothing ever stopped him, which was fortunate, because scientist-astronauts were firmly anchored at the bottom of the pecking order.

Jack was the only scientist in the program with a legitimate case for flying to the moon but fighting the pecking order wouldn't get him there. He became the unsung hero of one lunar landing mission after another, working tirelessly in Houston or at the Cape to prepare every crew from Apollo 11 on for their geological tasks.

As the Apollo program was drawing to a close, Jack finally got a break. He was placed on Gordon's backup crew for Apollo 15 even though it appeared to be too little and too late. Still, many of us felt the program would never end without Geologist Jack getting to the moon. He deserved it, and in some quarters he was considered the strongest item going for Dick Gordon in the competition for the last Apollo mission.

Gordon lost, but Jack did indeed get his flight. The real loser was Joe Engle, who had worked with Gene Cernan for a solid year as his Lunar Module pilot. Engle was replaced by Schmitt on the crew to fly Apollo

17, in the only instance of violating the pecking order that comes to mind (although Schmitt was selected nearly a year before Engle). It surely reflected the pressure the National Academy of Sciences put on NASA. Engle took it without a whimper. I'm sure I wouldn't have been as gracious.

Jack Schmitt represented a victory for all of the scientists who stayed with the astronaut program over the years, many with no visible rewards. They usually got the worst details, but they stuck it out.

Of course, underlying all of our actions and efforts and ambitions was gamesmanship. If the astronauts didn't invent it, we sure perfected it as an art form. It wasn't simply a case of knock-your-buddy, it was also a matter of knowing when and how to "boost" him, a form of damnation by faint praise: "Sam is smart. Sam knows his limitations and weaknesses." Some played the role of the sycophant while others were expert at the pointed barb.

Of course, in the Astronaut Office, no one was ever really secure. A space flight was such an overpowering goal that when one was at stake most of the astros became almost paranoid about their opportunities. The prevalent attitude was never to volunteer a favorable word about the way someone else was doing his job if he was even remotely considered competition. Of course, he had to be knocked *only* in front of the people who counted: Deke or Al Shepard or Chris Kraft, or any of the heavyweights up to and including Dr. Gilruth. Most important was any prospective crew commander, just in case they had any say about who they would take with them.

Everybody was fair game in this shoot-down-the-other-guy routine, but as a practical matter it was applied most effectively with respect to our peers, those at roughly the same place in the pecking order. Recalling my own last years with NASA, those equal to or above me in the pecking order received praise from me only on rare occasions, but I was never reluctant to praise a job well done by those who worked for me. In terms of the system, the latter was a harmless and acceptable practice.

We never were able to spell out the process by which one made it to a prime crew assignment, but we concluded that there was no science to it. Why Deke assigned Gene Cernan to the Gemini 9 backup crew with Tom Stafford remains a mystery, although the success of his subsequent career is not. Every rookie who has flown with a veteran, and performed adequately, could be expected to be looked after by his spacecraft commander. The only limitation to that rule was how well the commander's own career flourished and the failure of the former rookie to perform in the future.

This process was enhanced by the fact that crew selections were dependent in many ways on availability. And if you flew with one person, or served on a backup crew with him, you were both available in the same time frame for later assignments.

Tom Stafford's career was launched by an early and impressive performance on Gemini 6 with Schirra, and Cernan went with Tom on several backup crews, rode into space with him on Gemini 9 and Apollo 10, and finally headed his own mission on Apollo 17. When Tom was asked once why he had stuck with and, in effect, sponsored Cernan for so many years, he replied, "Gene knows how to follow orders. He'll do exactly what I tell him to do."

Alan Bean started off like many of us, believing that competence, dedication, hard work, and attention to the job would lead to its own reward: a space flight. It added up to being conscientious, but too much of it could be your downfall. If a man really believed in what he was doing, he didn't hesitate to express his opinion in reviews and meetings, even if it differed from those held by the higher-ups, who might not be as knowledgeable on the subject. It might be on a rule change in the use of the T-38, or whether the Astronaut Office should take a position different from the Program Office on a proposed spacecraft change.

All that the idealistic (early) Al Bean got for his efforts was a backup position on Gemini 10 when Elliott See and Charlie Bassett were killed, and heading the Apollo Applications (Skylab) effort when it was still seven or eight years away from flight. But Al was better at watching and learning than some of us, and apparently more pliable to boot. When he once decided to become "practical," he began to gradually improve his position.

It didn't happen overnight. He didn't just knock on Deke's door, and say, "I surrender, mold me," and bingo, become a company man. But Al began to swallow his own opinion, went along with the majority, no longer volunteered compliments about his colleagues, and learned to play the game. That wasn't an accident, but a conscious effort on Al's part—if you couldn't lick 'em, join 'em—and he recommended that I do the same.

With that program he became an accepted member of the establishment, was assigned to back up Apollo 9, landed on the moon on Apollo 12, and commanded the second Skylab mission for fifty-nine days in space.

In my own case, after flying the first Apollo mission I was assigned the job of chief of the Skylab Branch of the Astronaut Office, replacing Owen Garriott, the senior scientist-astronaut. Alan Bean and Gordon Cooper had taken whacks at it before Owen.

Two years later I was bumped by Pete Conrad, coming off Apollo

12. Pete was a second-generation astro who, in addition to pecking-order position, was well qualified for just about anything. The knowledge that Pete had the tools only softened the blow for me. I had expected to fly the first Skylab mission, having at one time been promised the assignment. When Pete took over he wanted the same thing and began looking forward to it. Then came the day Stafford said that if he decided against running for the Senate from Oklahoma he (Stafford) would displace Pete as the first Skylab commander. It was musical chairs all right, but some of the players were also starting and stopping the music.

In thinking back over the early years, a few obvious commandments come to mind, a sort of buyer's guide to how astropolitics was played:

1. If you were one of the boys you didn't necessarily have to be terribly competent.

2. You never worked for an astronaut selected after you, regardless of other qualifications—his or yours.

3. Physical fitness, other than injuries or death, had no bearing on your status, opportunities, or accomplishments.

4. First impressions, even those formed before you joined the program, could affect your career almost forever.

5. In bestowing rewards in the program, the rich got richer and the poor got poorer.

6. Scientist-astronauts stayed at the bottom of the pecking order.

7. Classical political influence—meaning out-of-town political influence—was counterproductive.

8. A power play was good only if you were sure you had the horses to finish it. That's why so few tried.

9. If you can't say anything good about someone, don't hesitate.

10. Do unto others as they would do unto you, but do it first.

7

Countdown to Apollo 7

Our stint as the Apollo 1 backup crew lasted barely two months
—from the Apollo 2 cancellation in late November 1966 to the night
of the Apollo 1 fire on January 27, 1967. It was a short but uniquely
different experience from being in the prime spot. In theory, the
backup crew is subject at any time to being placed in the shoes of the
prime crew, individually or collectively, and fly the mission. As a
practical matter, this is seldom the case and in fact has occurred only
two or three times in fifteen years. When we were thrown into the
breach to back up Grissom's crew it was without any thought of our
being called upon to replace them. There was serious doubt as to
whether even the prime crew would be trained well enough to fly the
mission at the scheduled date. To be frank, we were counting on the
schedule slipping due to hardware or testing programs which would
give them additional time to complete their training.

It all became academic that night in January when the fire on Pad 34
triggered a reappraisal of the entire Apollo program. The investigation
wasn't even complete when Wally, Donn, and I were notified we
would fly the next manned mission. Over the next several months
unmanned testing continued in the Apollo program and the mission
sequence was renumbered to account for them. Our flight would be
Apollo 7. Too bad we did not believe in luck.

By the summer of 1967 we were up to our ears in what was essentially a cowboy operation. We were trying to catch the Russians but

they were having problems of their own. Cosmonaut Vladimir Komarov had died during reentry in April 1967, but we still suspected (or claimed) they were ahead of the U.S.-proclaimed race to the moon.

The entire country seemed to be in the clutch of an emotional losing streak, of which the Apollo 1 pad fire had been a low point. The year 1967 had seen the war in Vietnam intensify. That summer the worst race riot in the nation's history ended in Detroit, Michigan, with 43 dead and 5,000 homeless. College campuses were seething. And in New York the schoolteachers went on strike for eighteen days.

The space program was getting a fresh look, from within as well as without. NASA was leaning hard on the contractors, demanding ultra-high levels of quality control and new safety practices. We had decided to bite the bullet; to take the inadequate junk, represented by the blackened pyre on Pad 34, and transform it into the brilliant machine which carried us to the moon in five giant steps.

North American Rockwell transferred Harrison Storms and brought in a new team to run its Space Division, while NASA replaced Apollo Program manager Joe Shea with a new man of their own—George Low. The program was desperately in need of new faces and a new approach. Low was an old-timer with NASA and had held many top management jobs. Both were young and bright but there the similarity ended. While Joe Shea was colorful and somewhat flamboyant, George was serious and businesslike at all times. Joe was aggressive and doggedly pursued the solution to problems as *he* saw them, including generating support for those solutions. George would thoughtfully consider the views brought to him by his deputies before selecting the appropriate course of action.

Partly because he was a contemporary, Joe was always willing to take on an astro at his own game, whether it was squash, handball, or turning a terrible pun. (His punning duels with Wally Schirra were legendary. Wally would attempt to overwhelm Joe with quantity, while Joe would rely on quality to carry the day.)

George preferred to remain aloof and above the fray, operating in the arena of mutual respect rather than friendly camaraderie. In George's decision-making processes, logic was seldom deterred by emotional appeal. He consistently placed good men in the top jobs and knew when to go along with them.

In that year and in that climate Wally, Donn, and I were at center stage. It wasn't a drill, a redundant mission, or a backup role. As the public saw it, the last crew had been killed, Apollo was still trying to get off the pad, and we—the crew of Apollo 7—were the men entrusted with *doing* it, with getting us back into the game.

My whole mood changed—the confidence in my approach to problems, my daily rhythm. I slept better. Food tasted better. My wife looked better. For the first time around the office our crew was being envied instead of envious.

The three of us were drawn closer by the irresistible pull of a shared adventure. Wally became, in every sense, the charismatic, complicated guy the press had been creating for most of a decade. It was *his* flight now; he wasn't standing in for Deke Slayton or backing up Gus Grissom. Until our recent crew announcement, when the virus really got to him, I honestly didn't feel that Wally had any urgent hankering to make another space voyage. His mind, if not his heart, was on getting out and exploring the world of commercial enterprise that would be waiting for him. As Wally himself put it, "The space age . . . devours people. I have been completely devoured by this business."

But there was one crucial circumstance that overlay his attitude—and ours. Wally Schirra was being pictured as the man who was going to rescue the manned space program. And that was a task worthy of Wally's interest.

Instead of just going through the motions, he became the wagonmaster. He was going to steer us and he left no doubt about it, even in those small moments when no one else really cared who was steering.

Wally could afford to be a prima donna on occasion, because he had one precious asset going for him: people adored him. He was simply a great, convivial, sunshiny guy. When Wally walked through a plant and shook hands with people, they had the dazed glow of people touched by royalty. Donn and I walked through and they were warm, friendly, and respectful, but they were just filing our names for future reference.

Schirra was a right-now hero. He could be ambitious, serious, contentious. He could be anything he wanted to be, including hungry or sleepy. If testing became a little hectic and Wally was not quite satisfied with a particular system, he would simply stop in the middle of a sentence and tell the engineers and the technicians to check it out one more time. "I'll be in the ready room, taking a nap," he would say. "Call me when it's ready."

Wally's and the program manager's attitudes weren't the only ones changed. Contractors became more receptive, to the point where we worried they would become too responsive to the sometimes arbitrary demands.

Between the night of the fire and the date of our launch twenty months and two weeks would pass. During that time the attitudes of the people at North American Rockwell and in the news media, as well as strangers in the all-night diner, conveyed to us the feeling that we were something special, like a Roman gladiator or the heavyweight boxing champion getting ready to step into the ring. Except that instead of measuring the time in minutes, we had to measure it in months. It enabled us to go about our work without bustle, but with a sense of purpose.

That's a lot of time to be soaking up applause or calculating risks. We drove ourselves, at work and at play. An airplane was always available to slip us away for a weekend of hunting on Catalina Island, or get us to Indianapolis on May 30 for a ride in the pace car at the Indy 500.

We could not afford to be invisible. The public had received a jolting reminder that there was a down side to space exploration, and it was our way—the program's way—of letting the country know we weren't running scared.

Between the weeks we lived at North American Rockwell to get the vehicle ready and the long hours we spent in the simulators, we probably qualified as the best-prepared crew ever sent on the first flight of a new vehicle.

With all that time, we still used it up and needed it all. As it was with every flight, but especially a first flight, they piled it on us. Of course, no matter how many more months we could have trained, there would still have been points not fully covered and details left undone. It was like getting married. If one waited until conditions were ideal, no one ever would get married.

The very best of times were those spent in residence at the Cape. It was a casual atmosphere, slacks-and-a-Ban-Lon-shirt, in the midst of hard work. But there was something about being on the scene that lifted our spirits. To keep it loose we met as often as possible at a private beach house NASA provided, taking in the sun and working on the flight plan. Life was a soft summer breeze.

We probably had more offices than Onassis. We kept one in Houston, one at Downey, California, and two at the Cape (one in the training building and the other in the crew quarters). We spent time in all of them, but we also made it a point to break the desk cycle and get away to smell the flowers and hear the music. This was part of the Wally Schirra philosophy: "Take an even strain." He didn't believe in burning yourself out. He thought it was important to cleanse the brain now and then and have fun.

The Cape was party-oriented, anyway. We lived it up and had our good times, usually with friends-of-the-family types. However much we partied, people defended our right to raise hell. They assumed it was an escape from the dangers soon to be faced. That aspect of the mission, the potential for death, was wildly overdramatized. Naturally, we enjoyed every minute of it.

We were all playing Mr. Cool, both because it would be poor form not to and because we really felt that way.

In the interest of maintaining our mental health, we lived in town, at the Hilton or the Holiday Inn, coming and going as we pleased, but taking many of our meals in the crew quarters.

Part of our continuing effort to keep ourselves amused centered around the infamous "gotcha," a simple shorthand for a form of practical joking, a game of personal tag. Wally would walk a long way and stay awfully late to pull a gotcha. Sooner or later we all got into the act, and I was involved in what was surely one of the best.

First, you should know that the crew quarters at the Cape was the only place where astros were authorized to call home at government expense. The telephone was a way of life to us. Even if we planned to go out and party on a given night—in fact, *especially* if we planned to go out and party—we felt better for having talked with our wives and kids first.

We'd place those calls through the NASA operators and over a period of time came to know each other's voices. The most famous voice of all was that of Operator 33. She not only had a sexy, throaty voice, she also had a sexy line and, well aware that she was talking to men who had been away from home a while, she laid it on thick.

Over the years, a remarkable number of us were suckered into impulsively asking, "Hey, how about having a drink after work?" The bears coming to the honey.

It would be a while before those who had met Operator 33 would confide in anybody else what their experience had been, but the word got around. She was not bad-looking, but for all her sensual talk, she was about as sexy as Martha Raye in army fatigues. In the flesh she was cold and painfully shy. It was all very depressing.

There was always somebody slow to get the word who would fall for Operator 33's "professional" voice. The sucker would end up at her place having a cup of coffee and sneaking glances at his watch, listening to her say something like, "Well, you probably want to leave now." And he would feel obligated to reply, "Oh, no, I'll have another cup of coffee."

It would be days before the embarrassment and guilt would wear off enough for him to laugh about it.

One of the men who hadn't been clued in was a young astronaut named Ed Givens, who happened to reach Operator 33 several times in a row during his calls home. We could read the signs, and so at night around the dinner table the rest of us laid it on heavy about what a hot number Operator 33 was.

Ed sidled over one night and asked, "Say, Walt, do you know Operator Thirty-three?"

I said, "Boy, she's a doll, isn't she? Don't miss out on that, Ed."

A couple of nights later all of us but Ed were having dinner together. Everyone had helped set the trap, and we figured, "Tonight's the night."

Someone said, "Ed's in cleaning up and changing clothes." A few moments later, Ed came down the hall that passed right by the dining room.

"Where you going, Ed?"

He stuck his head in the doorway and said, "Well, I have an appointment in town."

Someone said, "Oh, you got a date, huh?"

He really didn't want to admit a thing, but he looked right at me and said, "You better not be lying to me, Walt."

He walked out and we sat there, nearly choking with laughter.

The next morning we were at breakfast when Ed walked in, his eyes blazing. He ate breakfast in silence. Then, looking right at me, he had one comment: "You cheap sack of shit. . . ."

We all broke up again. I lived in dread of what he was going to pull to get even, right till the day Ed died in a car accident one year later.

As the months sped by, our investment in the job grew. Each day we gave another small piece of ourselves, making the prospect of a slip-up, of losing the flight, unthinkable. We would not *allow* some vagary of fate to knock us out of it now.

In the spring of 1968 we took my son Brian to see an orthopedic specialist about a foot problem, and I seized the opportunity to consult the doctor quietly about a worry of my own. I had developed a soreness in my left shoulder, presumably from a recent handball misadventure. The pain had persisted but I dared not visit the NASA physicians and run the risk of being removed from the flight crew six months before launch. After four and a half years, I wasn't going to blow it because of an aching shoulder.

Withholding medical information might not have been strictly kosher, but it was far from unusual. The determination to get into space carried us beyond the point of the usual anxieties about physical discomforts. They weren't so much health worries as they were career survival issues. Every time a NASA doctor poked around he had the power to prevent you from making your flight.

My conscience didn't even wince when I sought out civilian advice on what I hoped would be a minor condition. I regarded the NASA doctors with suspicion in spite of their record of finding some way to certify just about everyone for flight. But I felt it essential to protect myself from them and any risk of a shoot-from-the-hip decision.

Also, the NASA medics had acquired a poor reputation among the flight crews in their respect for the confidentiality of the doctor-patient relationship. Too often they had blabbed to the news media about this or that crewman's problem. We felt it was done for the notoriety, for a moment in the limelight at the expense of what should have been a private and confidential matter.

The doctor took an X-ray of my sore shoulder and after studying the picture insisted on taking several more. Then it was my turn to insist on a look at them. From the time we examined the complete set of X-rays, we began to play a little game with each other. He was concerned, I knew, about the profusion of small blemishes, or specks, on the bone itself that the pictures showed. Well, I was no doctor, but I had been exposed to a lot of medicine, and it took me only a few seconds to realize that he was concerned about bone cancer. During the next ninety days, while he checked it out, he never once indicated more than the possibility of an osteomyelitic condition.

He was my friend as well as my doctor, and he was willing to keep our secret. He kept insisting that he didn't believe it was anything to be unduly disturbed about but, in the way of doctors, thought it only prudent to get the most competent opinion. He proceeded to ship the X-rays to three different experts in three different corners of the country. I went home thinking that I was probably living with bone cancer and would not be around too much longer.

Two thoughts were uppermost in my mind, at least one of which, I'm sure, is common to anyone who suspects he has a fatal disease. I resolved to get my family affairs in better shape. And I hoped with all my heart that the NASA doctors would not discover the condition until after the flight of Apollo 7. Unless it somehow manifested itself externally, I felt the chances of concealing my problem were excellent. The flight, after all, was only six months away.

Three of those months passed in a kind of controlled suspense while the doctor and I waited for the judgments of the three specialists. I gave little thought to dying. It wasn't the first time in my thirty-six years that the prospect was near at hand, but it had always been in the abstract.

Living can be difficult, and three out of three people die, but there is no use railing against either. No matter how we juggle our standards of time, there are only so many hours in a day and so many days in a year. I have always believed the key is to live each of those days one by one . . . in doing not only those things we enjoy, but also enjoying those things that are done because we must. Mastering the things that must be done, as well as those you choose, will in time insure they become nearly one and the same. It's not how long you live, but how you live your life that counts.

While training continued and the doctor had no news for me about the X-rays, I had neither the time nor disposition to pity myself. Mine had been a full life spent in learning and doing. I could accept death, but it was going to make me mad as hell if I missed that flight.

Eventually, the doctor called me to his office. The findings were in. The diagnosis favorable. The condition turned out to be one peculiar to my own system. There were no other indications and I knew the bone cancer scare was behind me. I breathed easy again and, in time, even the soreness in my shoulder worked itself out.

I suppose your average fighter pilot sounds almost paranoid when it comes to his health. An aviator lives in terror of being grounded. A quirk, or a physician's pet peeve, can sometimes do you in.

Once I was examined by a navy flight surgeon at Los Alamitos, California. His hang-up was flat feet, and mine are so flat they can pass for rocking chair rungs. I had been tipped off about this particular doctor, and for the three years he was there I walked around his office with my toes digging into the floor to keep my arches elevated. All the while, I was aware the doctor was watching me out of the corners of his eyes. He suspected my feet were flat, which could disqualify a man for flight orders—because the regulations said so, that's why. And all this after I had been flying for twelve years. Once or twice he started to say something, but always held back. It chills me to think how easily my flying days could have ended right there. And space would have been something I'd look for in a parking lot.

As 1968 turned into autumn, the thoughts uppermost in the crew's minds were: don't screw up and don't get hurt and dropped off the flight.

The atmosphere grew more hectic as launch day approached. It was the natural consequence of the schedules drawing to a close, with all the last-minute details and the round-the-clock checkouts of the spacecraft. It was, after all, the first model of a new generation of space vehicles, and it was being checked out by a new contractor whose only previous experience resulted in a spacecraft being incinerated on the pad thirty days before launch. Finally, there was a flight crew that had been on the line for nearly three years, a crew that could have easily rebelled at the increasing flood of data being fed into the computer we literally wore around our shoulders.

Particularly irksome were the last-minute additions to the flight plan and the flood of input from the members of the hand-wringing society. Everyone wanted to make the mission safe, but they also wanted it loaded to make up for the long delay. Hand wringers are those folks who seem to come out of the woodwork, in the last few weeks before launch, with their own private concerns and reservations.

From the mission engineer: "On second thought, maybe you'd better not try that test with the computer—it might dump the REFSMAT."

A test engineer at the Rockwell plant: "We can't find in the records any mention of the overpressure on the water system last spring. What do you think about the following test? . . ."

A suit technician: "That suit will be OK, but if the zipper gives you any trouble. . . ."

The closer to launch, the more trivial and (it seemed) the more frequent we faced what Wally called FLT's: Funny Little Things. The expression dates back to an early geology field trip with C. C. Williams. During the exercise C. C. would occasionally bend over, pick up a rock for his pack, and say, "There's another FLR."

Like any well-trained astronaut, Wally was not prepared to show his ignorance easily. But as the day wore on, and the collection grew larger, he could no longer contain his curiosity, "What the hell is an FLR?"

"Funny Looking Rock," came the reply.

We began to lose patience with the FLT's, but we knew it wouldn't end till lift-off. Everyone wanted his conscience clean when the bolts blew. The showdown on flight plan changes wouldn't come till we were actually in orbit, and it would lead to some tense moments between Schirra and the control center.

The training became more realistic, and simulations included Mission Control and the other sites around the world. We worked in our pressure suits several days a week. The main objective in going through so many suited training exercises was to become familiar and comfortable in the spacesuit. It wasn't simply a suit in the usual sense, and the physical problems of living in it, especially pressurized, were like a frog learning to live in a turtle shell. Airplanes, boats, and other complex equipment can be bought for a whole lot less than the $140,000-a-suit price tag. In that mechanical marvel the simple function of taking a leak could become an imposing challenge.

Since we might be required to live in the suit for as long as four or five days, we routinely trained in it for long hours at a time. Provisions were incorporated to handle the "calls of nature" during the twelve-to-fourteen-hour days we spent lying on our backs for spacecraft tests or mission simulations. In shirtsleeves it was no trouble to make a head call every couple of hours, but for suited exercises it was virtually impossible. At those times we were damn glad they had developed a urine collection bag and a means of using it in the suit.

We attached ourselves to the collection bag through a piece of surgical tubing resembling a condom, a hose, and a one-way check valve. The first obstacle to overcome was the psychological block of "doing it" lying flat on your back. The second was overcoming the resistance of the check valve while urinating. The physical problem was sometimes easier to deal with than the psychological. Both could be overcome with practice, but even that didn't guarantee continued success. More than one astronaut was christened a "wetback" by putting off going because of the physical or mental block, eventually going out of necessity, and finally getting the waterworks turned on only to discover too late that the essential connection had come loose. Result: No pressure to overcome, no collector, and a warm, wet feeling around your middle. But it was far better to lose your cherry and become a wetback during rehearsal than it was in zero gravity.

For relaxation we had our gym: handball, tennis, and any other way we could find to unwind. Wally was determined to see that we stayed loose, and despite everything, we were probably the most relaxed crew ever to come down the pike.

Astronauts were notorious for using the Cape as their own private dragstrip. Dating back to the early Mercury days, the fellows had many an interesting shoulder-of-the-road conversation with the highway patrolmen. Usually, these encounters ended with the cop asking for an autograph and letting the offender go promising, as we always did, "to keep an eye on it next time."

In the event they were unrecognized, some of the astros felt no compunction about identifying themselves, and leaving the impression that they were in a hurry to catch the next rocket, or had an urgent appointment with the president.

I don't mean to excuse this blatant indifference to the traffic laws. But it was an attitude that spilled over from our hours in the cockpit, where one simply felt this sense of freedom, of privilege, of answering to nobody (almost).

About a month before the launch of Seven, Wally and I spent the afternoon on the public tennis courts at Cocoa Beach. Expressing his concern that I might injure myself at handball, Wally had leaned on me to take up tennis, *his* game.

We were returning to the Cape in two cars—Wally in his Corvette, me in a rented Oldsmobile—when the inevitable happened: a spontaneous drag race. Wally floored it, gunning away almost before I realized what he was doing. I'd be damned if I was going to let his Corvette walk off and leave me (the code again). I tore after him and spun around a curve in time to see smoke coming from all four of Wally's brakes.

He had screeched to a halt, for a reason that soon became painfully clear. Across the street, heading in the opposite direction, a police car was pulled over, and by the door a local gendarme was waving frantically for Wally and me to stop.

Coming to a stop behind Schirra I adopted my best meet-the-cop pose. We were both in tennis togs. The officer was finishing a ticket for some other luckless citizen, and I could see Wally fidgeting in his car ahead of me. His brain at that moment was working feverishly, trying to dream up a story that would melt that officer's heart. We're late for launch. There's a call on the Moscow hot line. The public commissioner is expecting us for tennis.

Impatiently, Wally started out of his car, confident that two reasonable men could dispose of the matter quickly. The cop spotted Wally making his move and angrily motioned him back to the car. Finally, he walked over, circled the Corvette, and began writing a ticket. Wally realized in about ten seconds that the cop either didn't know we were astronauts or didn't give a damn.

Slyly, Wally started working a little space patter into the conversation. The officer kept on writing. Wally bore in a little harder and dropped the fact that we were flying the next mission. "Yes," he said, smiling at the cop as if he was sharing a secret, "we're flying in another month. It's a great feeling."

It wasn't selling. I was growing more embarrassed and trying hard to remain quiet, tugging discreetly on Wally's shirt-sleeve. At last the cop had all he could take. "OK, both of you," he said, with a wave of the hand throwing me into the same pile of trouble. "Get into your cars and follow me down to the station."

Now, Wally decided discretion was the better part of valor. Now he clammed up, even though he felt we had been arrested unfairly. Oh, we were speeding all right, but there was no way for the officer, facing the other direction, to know by how much. All he had to go on was the dust we kicked up on the side of the road and the vapor trails we were leaving behind us.

We were ticketed and released, with a court appearance scheduled for one week before the flight. For just an instant I had a mental flash of the launch being cancelled and headlines all over the country screaming the news: "Apollo 7 Scrubbed; Astronauts in Can."

But Wally knew what to do. He called Charlie Buckley, the head of security at NASA, and asked him to "take care of it." Charlie, an understanding fellow and a racing buff himself, arranged for the court appearance to be delayed a week, which meant that it would have coincided with our second or third day in orbit. Not so good.

After a series of back-and-forth calls between Wally and Buckley—

since he got us into it, I was happy to let Wally try to get us out of it—our day in court was put off until after the flight. To my knowledge, that was the last thought Wally ever gave to the matter.

But the ticket, just the idea of it, preyed on my conscience. I asked Charlie what was going to be done about it. He shrugged. "They're not going to drop it," he assured me. "You guys really ticked off that officer."

After the flight I called the judge who had been assigned the case and asked what was required of me (no mention of Wally—it was now every man for himself). The judge said, "Walt, why don't you stop in and see me and we'll talk about it."

That seemed like a good thing to do, so I hopped into a T-38 and flew from Houston to Cocoa Beach, where my first stop was the courthouse. The judge turned out to be a genuinely gracious man. He gave me a tour of the jail, which I went to some lengths to praise profusely. When we finished he said, "Well, do you think you learned a lesson from this?"

"Yes, sir," I said. "I sure have. I'll never get caught speeding in Cocoa Beach again."

"OK, then, I'll just tear it up." He pulled out my ticket and Wally's. "Tell your buddy," he said, "that I took care of his, too. Ask him to stop in and say hello when he's in town."

I promised him I would, but decided to wait and see how long it took Wally to bring it up. That was in 1968 and I'm still waiting.

There was one more brush with the law and one more medical crisis to go before the launch.

Lift-off was set for October 11, 1968. On the Saturday before, at Wally's urging, we took advantage of the local dove season to do a bit of bird shooting. With a hunting fanatic like Tony Broadway on the staff, arrangements were no problem.

The gang of us took off for the property of a Melbourne, Florida, dairy farmer. The party included Deke and Stu Roosa, the astro serving as blockhouse communicator. It was a good hunt. Actually, we could hardly miss. Our friends had taken the precaution of setting up feeders in the area we were to hunt. The birds came in droves, and Tony kept busy just driving around the field, picking up the dead birds and dropping off another can of cold beer.

Late in the afternoon we had visitors, confirming the uneasy feeling about all good things coming to an end. The game wardens arrived. They didn't find all the extra birds we had shot, but they did find the feeders and proceeded to round us up in the field. It appeared we were headed for the hoosegow until it was pointed out to the wardens that their raid had netted the entire crew of Apollo 7: Wally, Donn, and me.

As usual, the astronaut mystique saved the day. We were released in time to join the rest of the astronaut staff, our medical personnel, and the game wardens at the dairy in Melbourne, where a barbecue was laid out. The afternoon's adventure provided food for conversation to go with the beef, booze, and watermelon.

It was one of those evenings when time doesn't seem to exist. And then, suddenly, around 6:30 P.M. I began having a sharp, searing pain that traveled across my chest every few minutes. By 7:00 P.M. I was convinced I was having, or had already suffered, a heart attack.

No one has ever accused me of being a hypochondriac. On the contrary, it is all but impossible to get me to admit I am sick. Pills, including aspirin, are something I rarely ever take. But now, for the second time in six months, it seemed I was in the grip of physical disaster. From what I could tell of my symptoms, at thirty-six I was having a coronary. Two of our doctors were at the party, but I gave only fleeting consideration to calling on them for help. Hell, in two days they would be giving me my F-minus-four-day flight physical, and I'd be damned if I'd give them any extra information they could use against me.

My immediate dilemma was solved when Dee O'Hara, our flight nurse from the medical directorate, called out for an early ride back to Cocoa Beach. I quickly volunteered, figuring if the very worst happened on the way in, at least I'd have some medical assistance.

It was a forty-five-minute ride back to the crew quarters with sharp pains hammering at my chest every three or four minutes. After dropping Dee off at the Holiday Inn, I called a friend, Al Bishop, and asked him to arrange a discreet visit to a doctor. Bishop was close to many of the astronauts and their families and could be trusted. A contractor representative and wise PR man, Al knew his way around the Cape, understood the NASA operation, and didn't need to be told why I was resorting to such private channels.

We met at the doctor's office. After listening to my symptoms, he gave me a thorough examination. The pains had subsided somewhat and were less frequent, but they were still there. He could find no sign of a heart problem and ran through the usual possibilities: gas pains, muscle spasm, and stomach upset. If he couldn't find anything maybe NASA wouldn't either.

The doctor made his final diagnosis as we were walking out of his office. Without knowing any of the astronauts or being familiar with the prevailing psychology, he made what was for him a logical assumption: "You know, with that launch just five days away, and you going into space for the first time, all that tension you're under could be causing this reaction."

He couldn't have been further from the mark, but it was useless to argue with him. I haven't spoken to any layman who hasn't assumed we were all under terrific tension due to the harrowing experience facing us in a space launch. Such is simply not the case. In eight years of watching flight crews prepare for a mission, I have observed only one or two men with even the slightest case of uptightness.

I thanked the doctor and turned to thank Al Bishop. "Forget it," he said. "You're not the first guy in the program to ask me for help with a medical problem."

In the room that night I dropped off to sleep in spite of the periodic stabs of pain. The next day the pain was still there, but the sun was shining outside and Cocoa Beach glistened with a fresh, lemony look. "To hell with it," I said, and by noon I had joined Bishop and his family at water skiing. By late that evening the pain was gone, never to return. But I didn't feel comfortable, and I didn't relax until they had completed my *F*-minus-four-day physical Monday morning with nothing detectable on the electrocardiogram.

It was almost time now. All the waiting, the frustrations, the misadventures, the oddball things, the pointless worries—none of it mattered now. I could even see the humor in a mild confrontation I had had with Wally over my penchant for pushing myself. On September 28, the wire services sent the following news story across the country:

Cape Kennedy—Navy Captain Wally M. Schirra, Jr., who almost lost his seat on a Mercury flight in 1962 because of a water skiing accident, has cautioned one of his fellow Apollo 7 crew members against daredevil ski tactics.

Civilian astronaut Walter Cunningham has taken a couple of nasty falls in recent weeks while skiing near this spaceport. Schirra, commander of the three-man Apollo 7 ship, made it plain he wasn't happy about it.

"I've chewed him out a couple times," Schirra said in an interview. "He tried to ski as well as I do and I've skied for years. But Walt, as he always does everything, was trying to catch up in a matter of months. I said, 'Walt, you can be replaced. We spent three years getting ready for this. If you want to fool around water skiing, these last three weeks, go ahead. But just realize what you're blowing if you do it.' So he's got the story."

Cunningham, who describes himself as a "go-go-go type," said in a separate interview that he "has to go full throttle all the time on everything, on or off the job." He said he took up water skiing for exercise because Schirra had complained about his charge-ahead tactics on the handball court.

The phone calls, the letters, the press coverage—everything was building to a climax. In Dover, England, the Flat Earth Society warned us that the round ball we would see below us was just an optical illusion. For months, mail had been arriving in bundles from everywhere, from kids and parents and soldiers and scientists.

A lady in Charleston, Massachusetts, for reasons unknown, singled me out to send along this cheery prophecy: "When you three go up in space, the rocket is going to get on fire and you three will be scattered in dust. Dust thou art, and to dust thou shalt return."

All that last week, Apollo 7 was the major story in the nation's newspapers. A headline in the Miami *Herald* announced, "Three Astronauts Ready to Face Challenge Three Others Died For."

Then it was October 11, 1968. I opened my eyes at 5:50 A.M., ten minutes early, as I customarily do in anticipation of an early call. It had been a sound sleep since 10:15 the night before.

I was thinking, "This is it . . . today's the day." The brass ring was finally swinging my way. I was exhilarated. Not with the breathless feeling of facing your first dive from the ten-meter platform, of forcing yourself to leap because you've come that far, but with a feeling of soaring achievement that you are there because you've earned it. It was like a first flight in the newest supersonic jet, only a thousand times better. No man had ever flown an Apollo spacecraft, and none had ever ridden as powerful a rocket.

We weren't stepping into the pages of history books. I knew that, but we would be a footnote somewhere: "First manned Apollo mission in the program to land a man on the moon." Among our peers, we were stepping to the top of the pyramid and you could feel the envy. They looked at us—and saw themselves. I felt a little self-conscious to be looking forward to the morning so anxiously.

The few minutes I had to spare were spent talking to Lo at her motel. The family had been up since 5:00 A.M. in preparation for going out on the Banana River to watch the launch from a boat chartered by *Life* magazine. This was to insure that the magazine got the "exclusive" photos for which they had been paying since 1963.

I had remembered to send Lo a dozen roses. There really wasn't much that needed to be said, she knew what they meant. We talked about the winds and the weather. I was a little concerned about Brian and Kimberly's reaction to the lift-off, but Lo assured me they were doing fine. She wanted to hear about our stag dinner party the night before. The crowd had included some of our old personal friends from entertainers to business executives.

I told Lo I would call her in eleven days, keeping it casual, and rang off. By then Deke Slayton was there to wake us, and we began the unhurried ritual which would lead to standing on that swingarm outside the spacecraft at 8:35 A.M.

Our first stop after showering—our last for eleven days—was the doctor's office. Wally and I walked together while he reminisced about the same walk prior to his Gemini flight. Part of the preflight physical is the Babinsky test, a check for lesions in the upper motor nervous system—nerve damage. The bottom of the foot is scraped with a dull instrument, while the toes are observed. If the toes turn up it is an indication of nerve damage.

Wally had practiced the proper response to gain one of his famous gotchas on the doctors. When Wally's toes turned up on launch day the doctors figured he was ready for a hospital bed—not a launch pad. Wally waited until they were huddling in whispers before he broke out laughing. There was no business so serious or so important that Wally could resist the temptation of a gotcha.

We had already undergone our major physical four days earlier (the F-minus-4). On launch day we breezed through a "cold mirror" test (that's where they hold a mirror under your nose and if you fog it they know you're alive). The doctors needed only twenty minutes to draw blood, take a urine specimen, and check our hearts, throats, ears, and noses.

As Wally and I returned to the crew quarters for breakfast we passed Donn Eisele. He was late and had been for most of the week. Donn was never known as a hard charger, but in this last week he had even slowed down a bit; it was hard to get his attention.

We shared some concern about his physical condition, his sharpness. His performance on the bicycle ergometer (a device to measure the effect of exercise on the cardiovascular system) four days before had been poor enough to cause the doctors some hesitation. Wally had made a personal plea to keep him on our crew for launch. Surprisingly, his marginal physical condition had little effect on his adaptability to zero g.

We had the usual launch-day breakfast of steak and eggs prepared by Lew Hartzei, the "astronaut cook." Lew had been with us for years but his helpers came and went. Outsiders naturally assumed that he was some kind of gourmet chef since he had been selected to cook for the space idols. The truth was, Lew was hired between charter boats and had found a home with the astronauts. While his cooking was overrated, he was easy to get along with and catered to our idiosyncrasies. There were homemade cookies several times a week because he knew they were my favorite food.

A half-dozen guests shared our breakfast, including James Webb, who had retired the week before as head of NASA (and was actually viewing his first manned space launch). John Healey of North American Rockwell, and Fred Peters and Ken Kleinknecht of NASA, also joined us, along with Deke and two members of our support crew, Ron Evans and Bill Pogue. Wally and Donn socialized while I read the morning paper, with its glorious headline (in sky-is-falling type), "Go 7!"

The camaraderie and underplayed excitement of an adventure soon to begin made the mood warm and pleasant.

After breakfast we made our way to the third floor of the checkout building to start the suiting ritual. It has been compared to the ceremony at the altar under the bullring where the matador dons his *traje de luces*, "suit of lights." No suit of armor was ever more elegant than ours, or more splendid, or complex, or essential, or expensive.

Waiting for us were the suit technicians, complete professionals who had probably gotten less sleep than we had. They were up late performing that last bit of lubrication on moving parts and checking the suits. We were each assigned a technician—Clyde Teague, Jim Lewis, or Keith Walton—and an assistant to help, if necessary. There was one more visit to the john before being wedded to the launch day urine bag. The bags would be strapped to our waists until we could get out of the suit in orbit hours later. A *solid* waste problem wouldn't dare intrude on our sensibilities during the next twelve hours.

The count was moving well. It would take a major problem between 7:30 and 11:00 A.M. for the mission to be scrubbed.

The technicians attached and checked out the biomedical harness. This "rig" consisted of five EKG sensors, signal conditioners, and associated electronics. Wally had engaged in a months-long campaign to simplify the harness. Now, at launch day, he balked. "Too many sensors," he complained. "The leads are too short."

He was betting on his launch day leverage, but Deke took charge in his firm, quiet way, and everyone went with the equipment on hand. Deke and Wally agreed to iron it out after the flight. It was destined to be lost in the noise of subsequent events.

All of our communications checked out, and our helmets were locked in place. We began breathing 100 percent oxygen to wash the nitrogen from our blood to lessen the likelihood of bends in the event the cabin didn't hold pressure during boost. At this point our interest was focused on the suit's checking out every step of the way. Any big problems would mean switching to a backup suit (another $140,000) lying ready on the bench. I had logged forty hours in the suit I was wearing and had confidence in it. The pressure points were familiar. An undiscovered pressure point, in a suit rarely used, could become maddening after four or five hours in the couch.

Every flight crew has an air of preoccupation at this time. Their minds are on the critical events the count must pass. Clyde Teague can cite examples of crewmen needing to be reminded to step into the suit, of stumbling through suiting procedures they had smoothly run through a dozen times.

One of the papers carried a launch day picture of me reclining in a lounge chair with my eyes closed. It was to stay relaxed and loose, save energy, and not get sweated up in the suit. But seeing the picture, I realized, relaxed or not, I would still have looked it. Of course, we consciously tried to project that cool image. We could, with a quick sideways glance, catch the technicians watching us, checking to see if we were measuring up.

I recall wondering if the program managers ever worried about a crew member saying on launch day, "I've changed my mind. I'm not going!"

When we stepped into the hall to begin our walk to the crew transfer van, our co-workers in the building were lined up to shout their good wishes. They broke into applause as we moved toward the elevator. Sound was muffled inside our fifty-six-pound suits, but we waved to acknowledge their support. Many of them were apprehensive, thinking about the fire twenty-one months before.

The elevator stopped at the first floor and, instead of continuing on by foot, Wally had arranged for us to climb aboard a small freight dolly to be hauled down the long hallway to the transfer van. All the way down that hall I felt foolish; we looked like three huge white sacks of potatoes. There we were, America's space gladiators, being pulled to our destiny in a freight wagon.

We stepped out into a nice day (there was a stiff breeze blowing that would come back to haunt us in a few hours). The applause from the hundreds of spectators waiting outside the crew training quarters, waving, clapping, and holding up signs, was a great feeling. But I knew that in two months there would be a new play with new heroes.

They were there to send us off and also to wonder what made us tick. If they only knew that most of them could have been locked in a capsule at lift-off and flown through launch, orbit, and reentry, with little or no ill effect—if everything went according to the script.

We looked at the pretty secretaries, smiling and waving, and even spotted some new faces and shapely legs we hadn't noticed before. It was good for a few smiles on the way to the pad and it was reassuring to realize, in that moment of fine tension, our vital signs were normal.

With a good imagination, you could think of us as businessmen in an alien world, on our way to a different office, carrying a strange briefcase (the portable ventilator with its supply of pure oxygen).

A police escort led us from Merritt Island onto the Air Force Eastern Test Range—the Cape—and past more cheering, waving crowds. How can so many people resign themselves to vicarious experiences? Perhaps Thoreau answered it best: "The mass of men lead lives of quiet desperation."

In this parade there is time to reminisce. I thought of how Wally got more livable, if not lovable, the closer we got to launch. He really wanted to go. It meant something now to make that one more space flight, to become the only man to fly each of this country's three generations of spacecraft: Mercury, Gemini, and now Apollo.

We could all feel the country pulling for us. The publicity and excitement was at a much higher pitch than it would be for many subsequent Apollo missions—more than the later lunar landings would receive. The Cape hadn't experienced a manned launch in two years, and we were putting them back into the flying business.

We didn't dwell on this. We were too busy living it. We took it for granted that the VIP bleachers would be bulging with prime ministers, governors, and big names in industry and show business. It was the in place to be.

The van moved quickly over the eight miles of flat terrain to Pad 34. Merritt Island, where the Kennedy Space Center is located, is just that—an island, with the Indian River on the west and the Banana River on the east. We crossed the Banana River to get to Cape Kennedy, which is air force–controlled. The undeveloped areas on the combined 84,000 acres are covered with tropical vegetation. Much of the concern about a launch abort is related to the fact that the Cape vegetation is not uniform. A pad abort, or an abort just after launch, with unfavorable winds, can blow the spacecraft back to crash on sand, water, or hard palmetto stumps.

The van made one stop at the blockhouse to let out Deke Slayton and Charlie Buckley; it was their station for launch.

The blockhouse is a two-story structure of twelve-foot-thick reinforced concrete, with only a dome protruding just above the ground. It's a sobering thought, sitting in the spacecraft with one million pounds of fuel under your butt, to know that all that steel and concrete is there for the protection of the VIP's and launch crew just in case we blow up. Little things like that bring home the reason why everyone in the world doesn't really want our job, even though they might claim they do at a cocktail party.

In the blockhouse everyone is quite safe. There are two periscopes and numerous closed-circuit television cameras for close-up viewing of all critical areas. Stoney (Stu Roosa, that day) has a console there, as do all the Kennedy Space Center officials necessary to make "go" or "no-go" decisions about a launch.

Less than an hour before lift-off the five-foot-thick, solid-steel double doors of the blockhouse will be closed. Once closed, no one comes out until after that monster has cleared the umbilical tower or the launch is safely scrubbed.

The gantry at Pad 34 loomed majestically over the blockhouse, even though it was a quarter mile away. The van complement was reduced now to the driver and those necessary to place our 500 pounds of human payload into the spacecraft. There was a stark beauty to the now empty area in the half-mile ride around the perimeter, to the bottom of the launch umbilical tower. I savored the thought that there were many, at that moment, who envied me the way I had envied Alan Shepard seven years before. Al Shepard was undoubtedly among them.

The van pulled up to the base of the $147-million complex, spectacular in its bright orange corrosion-resistant trappings, and backed up to the elevator. Clyde Teague, Wally, and I, still carrying our portable ventilators, unloaded for the ride to the 220-foot level. Donn would follow in ten minutes. We pressed the elevator button, and—nothing happened! My heart leaped, and the old pulse rate did a somersault. We punched the button several times; finally, the elevator started its creaky ascent. It left us wondering about our abort posture and the launch mission rules—the elevator had to be standing by at the White Room level for use in an emergency escape during certain parts of the countdown. It could mean a launch delay.

The White Room is a tiny, enclosed cubicle at the end of the swing-arm providing protected entry to the spacecraft. The preferred crew-escape route for critical emergencies, after strapping in, is across the swing-arm, into the elevator, and down to the ground rescue vehicle— an armored-personnel carrier, staffed by a four-man crew, which would meet us at the base of the tower to whisk us to safety. The rescue crew has just one responsibility in a crisis: to charge in, toss the three of us inside the vehicle, hang breathing systems around our necks, and take off through the boondocks. They were never needed, but they did enjoy those rehearsals, while scaring us half out of our wits.

In thirty seconds the elevator delivered us to the spacecraft level; Wally was the first to step out and cross the swing-arm to the White Room. We were about to realize a dream that, for me, had been five years in the making, including fourteen months of training for the cancelled Apollo 2 and twenty-two months getting ready for Apollo 7. I would soon find out how Captain Ahab must have felt when he harpooned his white whale.

Wally, Donn, and I had been together three years as a crew. Hundreds of millions of dollars were being bet that we would succeed in taking the first of the five Apollo steps to the moon. I felt supremely confident in my knowledge of the spacecraft and my ability to perform.

In the White Room we were greeted by the man we affectionately called the "pad führer." Gunter Wendt, a German immigrant, served as the major domo of the White Room for manned launches dating back to the Mercury days. His authority was nearly absolute at the spacecraft level where he tolerated no nonsense, from the flight crew on down. This attitude endeared him to us.

He had handled ten Gemini launches for the McDonnell Company, before North American Rockwell brought in its own people to organize the Apollo program. All of the astronauts, but especially Wally, had pulled strings to persuade North American Rockwell to hire Gunter. Some of us believed that Gus Grissom's crew might still be alive if Gunter Wendt had been on the scene that day.

We considered him a key man. North American Rockwell was new in the manned spacecraft business. They had some good men, like Skip Chauvin, the test conductor at the Cape, and John Healey, the spacecraft manager. But they had yet to launch their first manned vehicle, so it was an added touch of confidence to have Gunter there.

While Wally was being strapped in, my portable ventilator suddenly began to lose oxygen pressure. With efficient NASA planning, there was a spare available in the White Room at the end of the swing-arm. It was soon apparent that I had to be switched over or suffocate. But the technician and I both hesitated. On our minds was the possibility of invalidating the lengthy checkout procedure. I had been literally certified, signed, sealed, and delivered. With a look between us, we went ahead and made the switch and I told them about it after the flight.

It was quite windy now. Beneath us we could feel the vehicle and swing-arm sway. I thought how often over the last seven years I had envied other men this moment, as they climbed to the top of other rockets and rode their way into the history books. As I stood on the swing-arm 220 feet above Pad 34, waiting for Wally Schirra to be strapped into his couch, the emotion just wouldn't come. The memory of Al Shepard's launch in May 1961 seemed more real than the launch day events for Apollo 7.

Looking down at that beautiful booster in the clouds of liquid oxygen, venting off excess pressure, bathing the American flag on its side in fine mist, I kept waiting for it to seem real. I had the eerie feeling that it was just another drill. When would it become a reality? Standing on that swing-arm for seven or eight minutes, in a wind that felt like a gale, it never came. Was it a characteristic of the business that I wasn't to experience the pleasure of fulfillment as we stepped through the well-rehearsed script?

While I stood there in the blowing wind, only a few short miles away on the Banana River, Lo, Brian, and Kimberly were watching that same oxygen venting in the morning sun. They were following launch activities on commercial television and keeping their fingers crossed. NASA had been less than enthusiastic about their being at the Cape in the first place and left us on our own for travel and accommodation arrangements. Ironically, they were prepared to help on the return trip—if the circumstances warranted. Prior to boarding the boat that morning, Lo was notified that NASA would have a car standing by at the dock to whisk the family to Patrick Air Force Base for a government flight back to Houston in the event of a launch catastrophe. It was phrased a little more delicately but that's what it amounted to. The offer was appreciated when I learned of it later, but on that morning Lo certainly did not want to think about it.

To an astronaut, the mission was to get off the ground. He enjoyed the fruits of everyone's labor, the glories, the glamor, and the rewards. But to the suit technician, the launch is strapping us into the spacecraft. It was their big scene in the months-long drama, as important to them as lift-off was to us.

Wally was strapped in and then it was my turn. While trying to fasten a stubborn shoulder harness, Clyde Teague cut his knuckles and worried about bleeding all over our white suits. At that stage we would not have held the count for even a cut throat, believe me.

Wally and I were able to talk to each other without mission control hearing by way of a communication mode we had ferreted out in the circuitry of our radio panel. Donn, who was by now strapped in, could not reach his switches and had to content himself with facial expressions or transmit to the world.

T minus two hours and counting: John Young, our backup Command Module pilot who had been in the cabin all morning going through final switch settings, left us to descend to the blockhouse. All around us people were climbing into bombproof shelters while we lay, just a little impatiently, on top of a million pounds of high-energy fuel.

Wind conditions remained the only questionable factor in our countdown.* The launch director didn't want to scrub or delay the launch for high winds, and neither did we. We knew that the abort criteria had been compromised to the limit for this mission on top of all the other unknowns it faced. For example, during the boost of our Saturn 1-B, there was a period of approximately ten seconds during which an abort would impose more than 18 g's on our bodies. It was the first time this exposure had been accepted, and then only after much soul-searching.

---

* We could tolerate winds up to 18 knots measured at the sixty-foot level.

T minus 1 hour, 9 minutes, and counting: Wally peered out the upward viewing window from the pad—he is the only one in a position to see the sky—and said, "It's as blue as a bluebird out there."

T minus 50 minutes, and counting: Everything is "go" for Apollo 7. The terminal count is picked up with the communication checks from Stoney (Stu Roosa). If we got into a prolonged hold, Stu could give us a direct phone connection with our families.

It was time for the last of the pad crew to clear out. As Gunter left the White Room, Wally could not resist a parting pun. "Gunter went," he announced to the blockhouse.

After the White Room was cleared out, the access arm was swung back three feet from the spacecraft. For the first time, we're able to arm our pad abort system and, if necessary, use the launch escape tower to lift the Command Module away from the booster. We are alone at last, and it is great!

It was turning out to be the smoothest final count ever for a first flight. The latest wind report was 15 to 18 knots from the east, with gusts to 22. Inside the spacecraft we are resigned to having an out-of-limit launch condition with the winds, but we aren't happy about it.

T minus 33 minutes, 30 seconds, and counting: We receive our first glitch. The high-speed elevator, which was to be stationed at the 220-foot level for our quick evacuation, was stuck at the ground level after taking down the close-out crew. If anything goes wrong, we will have to ride down the emergency slide wire.

We're still keeping an eye on the winds. To abort, we needed only to fire the 155,000-pound-thrust launch escape rocket, but that wouldn't get us above the fifty miles needed to qualify for astronaut wings. I wondered idly what Al Shepard thought about a launch escape rocket with more thrust than his Redstone booster?

T minus 25 minutes: Wally couldn't resist getting his doubts about the launch winds on record, putting the boys in the blockhouse on the spot. We were convinced the winds exceeded the launch mission rules, but everybody, including us, had go fever.

T minus 6 minutes, 15 seconds: There is a hold, a minor problem that cost us two minutes while the inner chamber of the second stage engine was being cooled to prescribed limits. The delay disturbed us out of all proportion to the time lost.

The count continued.

T minus 4 minutes, 7 seconds: Skip Chauvin, our launch conductor, said, "Go for launch." The destruct system for the second stage of the Saturn 1-B is activated, and we were standing by to go on the automatic sequencer. We all give a final "go" inside the spacecraft.

T minus 2 minutes, 43 seconds, and counting: From this point on, the countdown is automatic, with all launch functions initiated by the computer in the blockhouse. We knew we were going—finally! Each of us was fully occupied now, and even five or ten seconds of idleness seemed like an hour. The two stages of the Saturn 1-B were starting to pressurize with helium. The pressure was building up.

Two minutes and counting: There is little chatter on the circuits. Donn is completing the key check in the guidance and navigation system. The abort advisory system in the spacecraft is activated. Inside our gloves, fingers are crossed against any late problems.

Sixty seconds and counting: "We're go for Apollo Seven."

Fifty seconds: The vehicle is pressurized. "The spacecraft is go!"

Forty seconds and counting: All aspects of the mission are go. I can't resist thinking, "In forty seconds we will have the first fire at Pad Thirty-four since that ill-fated test of the Apollo One vehicle."

Thirty seconds. The booster is switched to full internal power.

Twenty-one seconds: The power transfer is complete. In ten seconds we will have 1.6 million pounds of thrust pushing up our tail.

Roosa makes the traditional count backwards from ten, until he gives the call that would send shivers up the spine of a Greek statue:

"One—we have ignition—we have lift-off."

8

The Flight of the Phoenix

THE ROCKET RISES from the pad so slowly, so ponderously at first, that you could be imagining it even moved at all. Painstakingly, it climbs, trailing a fireball as vivid as the colors of hell. From the spectators' level the earth actually trembles. The vibrations, the noise, the shock, roll over you in waves. From out of the *second* fire on Pad 34 rises our modern-day Phoenix.

I'd seen many lift-offs, but this one was different: I was riding inside, not watching.

Half a million Americans lined the roads and beaches of Florida that day to watch the United States reenter the business of manned space-flight. It isn't stretching the truth or overdramatizing to say we carried the nation's hopes with us. Twenty-one months before, a fire on the very pad from which we launched had killed three of our teammates. One more setback now, and the prospects of landing a man on the moon before 1970 would be gone forever.

In the cabin we had no way of sensing the moment of lift-off. We knew the Phoenix had started to climb from out of the fireball only because the spacecraft clock was ticking off our elapsed time and Wally could see the altimeter climbing. Donn had the computer flashing our trajectory parameters. I watched the environmental and electrical systems to insure that the cabin pressure was bleeding down according to schedule to prevent a blowup in the vacuum of space.

For the first two minutes, thirty-six seconds, our only visual contact with the outside world was through a small forward viewing window on Wally's side and a small round porthole over Donn's head. But then the launch escape tower hauls the boost protective cover away and all five windows are exposed for the first time. There was nothing to see but sky until the last few minutes of boost. Donn started rubbernecking soon after the cover cleared, and we both snapped at him to keep his mind on the job.

Apollo 7 was fifty miles high and sixty miles down-range when the word came from Houston, "Seven, you're looking good."

At four minutes after lift-off, Wally radioed back, "Apollo Seven is go." The country's first manned flight in nearly two years was off and running, and it was riding like a dream. A few moments later our CapCom in Houston, Jack Swigert, gave us a call: "You're right on the old button."

For the first ten minutes, except for performing a few essential functions, we were literally along for the ride. We're space pilots, but we aren't flying the spacecraft. And Mission Control doesn't control it. It was an on-board computer operation all the way. We were on automatic pilot. The flight path had been programmed. We watched the instruments—so did Houston—to see how good a job Seven was doing flying itself. If it didn't perform, we could either fix it or get off.

We might not have been piloting the craft, but we weren't exactly idle as we entered an eleven-day coast around the world at 17,500 mph. In those first hazy minutes, we had to run through the long insertion checklist, get our helmets off, retrieve some of our stowage, and—during the second pass—separate from the Saturn 4-B, the second stage of our booster.

The second time we passed over Africa there was a quick chance to look out and let the wonder of the moment sink in. I looked down and there was the Sinai Peninsula, just as it had appeared in drawings of the recent Arab-Israeli war, with the Suez Canal and the Red Sea on one side and the Gulf of Aqaba on the other. It was fantastic. Here I was, looking at the globe as it *really* was, and I thought, The early geographers, the people who made those first globes, were right all along. The camera was unstowed by now and I took some pictures and recorded the frame data.

Once the initial activity had subsided, my first sensation of space was one of simply belonging. We had lived with it for five years, had seen the parade of new actors coming onto the stage, dominating the scene, taking their curtain calls, and then scurrying away with greater notoriety and inflated self-confidence and fame. A few months later another cast, the same sudden exposure, the press conferences, the questions, and now, it was us!

A fellow would have to be less than human not to have his consciousness raised during those first hours. We had embarked on the very first of what would be literally hundreds of firsts for the Apollo program. The three most significant milestones, as we rated them, were:

1. The *first* landing on the moon.
2. The *first* circumlunar flight.
3. The *first* manned flight of the Apollo spacecraft.

There were significant events with every turn of the clock: the first American three-man flight; the first Apollo rendezvous; the first celestial navigation fixes; the first manned firing of the 20,000-pound-thrust service propulsion engine.

And it would all go so smoothly, so uneventfully, that any preflight doubts seemed almost foolish. At lift-off we weren't worrying about dying or having that 230-foot rocket blow up. The only worry I recall was the fear of failure. It showed itself in a little pep talk I delivered to myself aloud just before launch: "OK, Cunningham, whatever you do, *don't screw up.*"

I had always envied engineering test pilots with the aircraft companies, the men with an opportunity to take the first hand-built model of a brand-new aircraft into the air for its first flight. Now here I was in a piece of machinery many times more complex than any airplane ever built, carrying men for the first time into a sterile, strange, and uncharted sea. It was the first manned test of the most powerful booster in the world and its last use till the Skylab program five years later.

Since the last American space flight, the United States had lost a crew of three and the Russians had lost a cosmonaut in a space accident. Confidence needed to be restored that outer space and man belonged together. The public had to be convinced that our missions were not circus stunts performed by soldiers of fortune with a collective death wish.

And, of course, we were keeping an eye on the Soviets. In a shortsighted effort to keep our space budget up to par, the race to land the first men on the moon was pushed for all it was worth. The world came to regard it as a great Olympiad in the sky. When Russia launched Zond 5 in September 1968, we worried that it was a prelude to an imminent Russian manned flight around the moon and, eventually, a lunar landing. By the time Apollo 7 lifted off, we knew that Zond 5 had reentered the atmosphere in a trajectory—steep enough to heat the craft to levels that only instruments, not humans, could safely withstand. Our spirits soared. It was clear that Apollo 7 would put the United States back in the lead.

At home, North American Rockwell had been the prime Apollo contractor for more than ten years without launching a manned spaceship. They were looking to our crew to put them on the scoreboard. Beyond question, Apollo 7 was the most ambitious first flight ever undertaken. By contrast, John Glenn's first Mercury mission had lasted less than five hours and the first Gemini flight, with Gus Grissom and John Young, did no better. Apollo 7 was to be our second-longest space mission right out of the starting blocks. Its planned duration of eleven days would demonstrate the capability of all the on-board systems to sustain a crew for the time required to complete a lunar landing mission. This was our first step to the moon.

You don't really know what togetherness is until you have spent eleven days with two other people in a volume about the size of the back seat in a large American sedan. If your buddy belched, you said, "Excuse me." We worked, ate, slept, caught colds, and performed all bodily functions in that same space. Surprisingly, other than rare outbreaks of temper, it was less of a problem than one might have expected. We had spent as much as twenty hours a day in the spacecraft during testing and simulations, so we were prepared for possible psychological strains.

The one thing that made it more tolerable was the absence of gravity: living in a continual state of free fall. On earth we move essentially in a two-dimensional world. In a house, for example, we always compete for space with furniture, lamps, dogs, and other people. Think how much roomier it would be if you could avoid all those obstacles by walking on the ceiling or the walls, or doing as we did in the spacecraft —flying across the middle of the room.

("Flying" is somewhat of a misnomer for the Apollo spacecraft, because we could always extend an arm or a leg and touch the walls on one side or the other. In Skylab, however, it became a very apt description.)

In the world of zero gravity, volume is the measure that restricts movement and becomes confining, not the floor area. It is ironic that as man moved into the vastness of space his own movement became more restricted—to 225 cubic feet, the interior of the Apollo spacecraft.

What is zero gravity like? That's a difficult question to answer. Those who have experienced zero gravity find it hard to explain and those who haven't, can't even begin to imagine it. It is a condition which cannot be created artificially. It can be experienced imperfectly during parabolic flight in an airplane for a few seconds. But experiencing zero gravity, or free fall, is like pregnancy. There is no such thing as "almost."

A feeling which is amazingly close, is the sensation experienced during a dream of falling out of bed, or falling interminably through space, with nothing to break that fall. That "free falling" feeling just prior to awakening, with the terror stripped away, is the nearest thing to floating in zero gravity I have ever experienced.

Sight and feel, both conscious zero-gravity senses, adjust almost immediately. The subconscious is not nearly as flexible. It requires two to three days on first exposure to overcome that falling sensation immediately prior to sleep in the spacecraft.

Into that small but compelling world came the three of us. Twenty-four hours later, Wally was sick with a miserable head cold, followed the next day by Donn's, which was only a little less severe. I never caught the cold, although I felt blah the third day. (A week before the flight I had shown symptoms of a cold, and a couple of days later Donn and I took shots for it. Wally had rejected the idea since he felt fine; he couldn't imagine himself catching a cold at such a historic moment and wasn't about to get a shot just for the exercise.)

By the third day Wally was miserable. In our on-board storage we had included a large supply of vacuum-packed tissues. They were useful for wiping off the grime, catching urine drops, or any other of the many required body ablutions. We had fortunately overstocked on this item, but the used ones were not vacuum-packed, and took up approximately twice as much space.

As soon as we were squared away in orbit, Wally began blowing his nose, and within a short time we were stuffing used tissues in every unused spot we could find. Wally didn't seem interested in conserving them, and Donn and I began to worry whether a tissue crisis would end the mission. We hinted to Wally that he should try to average more than one blow per tissue, but our pleas went for naught. We were literally up to our asses in used tissues.

For a couple of days we experienced the space equivalent of staying home sick in bed, with Wally feeling too wretched some mornings to get out of his sleep restraint beneath the left-hand couch.

When Donn developed his runny nose the impression was left with Mission Control that the entire spacecraft was rife with colds, but actually only Wally had a real problem. It's the kind of thing Bill Anders referred to as "The Bull Moose Complex," from Al Capp's character, General Bull Moose: "What's good for Bull Moose is good for the world." Or, paraphrased: what was good for Wally was good for our world—the spacecraft.

There was no doubt that Wally felt he had caught the cold they were saving for Judas. He was sleeping poorly and becoming as jumpy as a frog in a biology lab. He continued to perform, even issuing orders from his "sickbed," but the high work load at the outset of the mission, combined with his discomfort, made Wally more irascible by the day. He didn't miss an opportunity to nail Mission Control to the wall. Donn and I were amazed at the patience of those in the control center with some of the outbursts that came their way. On the ground they were well aware that every word of the air-to-ground communications was being fed directly to the press center, a fact of which we had not been informed. So Wally's bad temper was making big news back home.

The first flare-up came on the second day in orbit with the first scheduled television broadcast from space. Due largely to his cold and some minor problems we were struggling to keep abreast of the schedule. "I tell you," he snapped, "this in-flight TV will be delayed without further discussion." That ended the debate, and ground control agreed to let that broadcast go till the following day.

That was all the encouragement Wally needed in the rigid stand he took throughout the mission. It was also the only time obstinance was warranted. We just plain weren't ready for a TV broadcast.

Wally's reason for canceling the early TV transmission was to concentrate on our rendezvous with the Saturn 4-B. He had reason to be a little uptight about it. It was being done without radar and the things that could go wrong were too numerous to count. We fired the main engine twice: once to put the capsule in a position below and slightly behind the Saturn 4-B and a second time to bring us close enough so that we could accomplish the final matchup. It went off without a wrinkle. Wally maneuvered to within 70 feet of the rocket and flew alongside, in perfect formation, until it was time to break away.

Eventually those daily television shows would become the entertainment highlight of the journey—for us as well as for the public. Lasting seven to eleven minutes, they came to be called "The Wally, Walt, and Donn Show." They were scheduled once each morning during a roughly 2,000-mile pass between Texas and Cape Kennedy, the only two ground stations equipped to pick up the transmissions. The ham in us didn't just surface; we damn near brought back vaudeville. We began the broadcasts with a sign that read, "Hello from the lovely Apollo Room, high atop everything." And once we displayed one that asked, "Deke Slayton, are you a turtle?"

According to barroom tradition, Deke was required to answer, "You bet your sweet ass I am," or pay the penalty of buying a drink for everyone within earshot. Deke copped out by cutting off his microphone and recording the answer.

The telecasts gave us a chance not only to check out the system that would capture the first footsteps on the moon, but also to demonstrate to the people at home a few facts of space life. We performed acrobatics at one point to provide a glimpse of weightlessness in action. We showed scenes out the spacecraft windows, catching glimpses of clouds and coastlines racing by. But our best show was done completely upside down on the home screens. It was an opportunity to have fun and strip away some of the mystery. The shows also led to a special Emmy Award from the Academy of Television Arts and Sciences the following year.

Wally grew more and more impatient with maneuvers and tests he felt were not part of the original plan. At one point he said, "I wish you would find out the idiot's name who thought up *this* test. I want to talk to him personally when I get back down." Then Donn Eisele got caught up in the swing of it and began to throw a little weight around. Donn's complaint had to do with a test to get a navigation fix on the horizon. "I think it's rather ill prepared and hastily conceived," he said. Echoing Wally, he added: "I want to talk to the man who thought up that little gem. That one really got to us."

Finally, Wally laid down "Schirra's Law": "I have had it up here today, and from now on I'm going to be an on-board flight director for flight plan updates. We are not going to accept any new games or do some crazy test we never heard of before."

We learned later that at the next press briefing, a newsman stood up and said to Glynn Lunney, the prime flight director for Apollo 7: "I've covered sixteen flights, man and boy, and I don't recall ever finding a bunch of people up there growling the way these guys are. Now you're either doing a bad job down here, or they're a bunch of malcontents. Which is it?"

"I would be a little hard pressed to answer that one," said Lunney, and dropped it.

The flight controllers in the trenches—the rows of consoles where the various spacecraft systems were monitored—tended not to pay a whole lot of attention to Wally's testy exchanges, or chose to look the other way and ignore the embarrassing situation they created. Lunney, like the other flight directors, was faced with that press conference every time the shift changed. He was on the hot seat with the media at least once a day. He later confided that he didn't feel it affected his judgment or performance, but during every slow period he caught himself dwelling on the effect Wally's indulgence was having on the professional procedures and rapport carefully built over the years. He felt it was so obvious to both astronauts and controllers that he speculated on whether Wally's buddies would take him out behind the woodshed and shape him up when he got back.

When Lunney felt things were really bad he began tapping his pencil; George Low, the Apollo program manager, would try to calm him down by talking about other things. Deke, who was himself highly incensed, contributed by telling jokes.

The backup crew and support crews working in Mission Control also suffered during the mission from defending Wally's actions in some pretty indefensible positions. For guys like Jack Swigert, Bill Pogue, and Ron Evans it may have had a carryover effect on their future crew assignments.

There's no question that it had a significant effect on the mission. Our flight directors were very reluctant to send up anything additional, new, or different in the way of tests or checkouts even though there was plenty of time toward the end of the mission for such additional tests. They felt that such additions would only present further opportunities for embarrassing exchanges on the air-to-ground communications or that some worthwhile test would not be understood by Wally, which would then lead to an outburst.

In retrospect, the exchanges weren't all that significant, but in a mission as routine and boring as Apollo 7 turned out to be the media was desperate for any angle.

Wally saw each encounter as a challenge to his authority and judgment as captain of the ship. He never felt he was behaving like an ass, but rather that he was defending a principle. My own attitude was cautious, if not downright cowardly. I did not feel justified in the behavior and abuse that was being heaped on the ground. I also realized that all three of us would be tarred by the same brush. So when the exchanges became heated, or threatened to, I would disengage myself, stay out, and later be the apologist for ground control's position. The new tests *were* sometimes poorly thought out and *did* add to our work load (which was not overloaded), but they also produced valuable data.

In the end the mission left a bitter residue with the support people and controllers who had worked with so many crews before. In Glynn Lunney we had the best flight director the Manned Spacecraft Center ever produced, and his patience was tried to the limits. The technical success we achieved on the mission was blemished to a great degree by the reputation we earned for being difficult. Then, too, the mission objectives were accomplished so routinely that the news media had very little to report except those "human interest" items.

In retrospect, at least four factors combined to make the flight a pressure cooker for Wally Schirra: his age (he was forty-five at the time); his preflight lack of interest in the details of many of the test objectives; the nuisance of his head cold, the effects of which were heightened by zero gravity; and the long duration of the mission. His longest previous stay in space had been twenty-five hours, and Wally was noted more for his quick style and grace than for his staying power.

When one considers the strange sleeping and eating cycles we endured, and the closeness of our quarters, it's remarkable that more tensions didn't result. On our first day in space we were up for twenty-three hours straight before getting to sleep. NASA liked to get its money's worth from every crew—and did, especially at the beginning of missions.

During training, I had fairly well sorted out what it was going to be like in a spacecraft for eleven days, including how I would behave and respond. Working with Wally and Donn for nearly three years—prime, backup, and prime crew again—I knew their strengths and weaknesses. If they were paying attention, they knew mine. I had anticipated certain problems and knew what I intended to do in such cases.

"I'm the navigator," Donn, our navigation expert, liked to joke, "and I've got a right to know where we are." But he knew the guidance and control system and had a good grasp of the mission and its objectives. Together we had developed the procedures to accomplish the mission objectives, and we had cross-trained enough, including flying reentries, that we were confident there was no aspect of the mission the two of us couldn't handle. If anything happened to Wally, we felt we could cover for him. It wouldn't surprise me to find out each of us had privately reached that same conclusion about each other's jobs.

As a practical manner, Donn's job did not revolve around life-or-death situations. The decisions that dealt with our very necks were Wally's. He was in command of the vehicle. It was his hand on the abort handle during boost; he flew the spacecraft and handled the attitude control system 90 percent of the time. A dumb mistake by any of us could blow the mission or, worse, the spacecraft. But if everything went according to Hoyle, as we expected it to, neither Donn nor I was in a position to cause or react to time-critical problems. Wally was in control at the crucial times when our necks were on the line and that is what he trained at.

The catch was that Wally hadn't exactly immersed himself in the other objectives of the mission prior to the flight. Donn and I suspected that some of those objectives would only sink in during the flight causing him to raise points, reservations, and complaints that should have been made months before. Development of the in-flight procedures had fallen to Donn and me by default.

Wally was never a detail man, and at least part of his testiness during the flight was caused by his discovery, for the first time, of some of the details of the tests we were to perform. Then there was the added stress of a bad head cold and the realization that if the mission had to be aborted, there was a high risk that his eardrums would be damaged in the subsequent reentry. It added up to Wally being an irascible fellow most of the trip.

It wasn't typical of me, but I decided to roll with the punches and catch as little flak as possible. That left me in several instances acting as a mediator between Wally and Donn. Now, I'm not noted for my diplomacy, but the only real on-board confrontation of the mission taxed what little I had.

Every mission had literally hundreds of test objectives to be accomplished and, invariably, one task runs right up against the next. On our ninth day, Donn was required to verify the midcourse navigation program in the Apollo guidance computer. This involved maneuvering the spacecraft to a prescribed attitude (requiring some expenditure of fuel). Then Donn, using the sextant and computer, was to track a star on the lunar horizon.

During the next night-pass, Wally was scheduled to check out a backup alignment system by maneuvering to several stars using a sight placed in one window. His effort would be expensive in terms of fuel expended, also, and therefore was scheduled to be performed only once during the mission.

Donn's task was going slowly—it had never been attempted even in the simulator—and Wally kept chipping at him to hurry. I tried to point out that we still had twenty minutes of darkness left, but Wally barked, "Wrap it up. I've got to do this backup alignment."

Donn had been working hard on the test and had just one more sighting to make. He finally got fed up and snapped back, which for him was uncharacteristic: "Okay, if you insist, I'll wrap it up. But we've already spent the fuel to get here and I could be finished in another five minutes. I just want to put it on the tape that we're knocking off this test because you ordered it."

Wally didn't even blink. "OK," he said. "You just do that and when you get through I'm going to put on that same tape that you were threatening mutiny."

I was over on my couch trying to laugh it off and stick a word in edgewise, as a peacemaker: "Well, Wally, why don't we wait a couple more minutes." And, "Donn, we can come back to it."

It all blew over rather quickly, with Donn making one more sighting. They spent a frosty day pouting and saying as little as possible to each other, and then the air just cleared. Donn knew there was no room for that kind of blood in the six-foot cockpit, and Wally, I suspect, must have felt a bit foolish about the whole affair. At any rate, not another word was ever mentioned again, during or after the flight. None of it ever appeared on tape, either.

Donn and I were both curious about the reason behind Wally's impatience with the navigation test. We assumed Wally had wanted to start a slow, fuel-conserving maneuver to line up his stars—a bad guess. Apparently, Wally had felt uneasy about performing his alignment—one try only—under the watchful monitoring of Mission Control on the next night-pass. He had decided to rehearse the entire process. With darkness running out, he had to really burn up the fuel to get his practice alignment completed. An hour later the real thing went off flawlessly, and to this day, ground control hasn't figured out what happened to the forty pounds of fuel that came up missing after the ninth day of operations.

What surprised me most about the episode was that for the first time in three years, Donn had stood up on his hind legs and said, "Just a damned minute, Mr. Schirra." It was something Donn should have done a hundred times in the past rather than laugh along with Wally and compromise a point in training. If he had, the two of us would probably have been able to force a better balance in the crew when we finally flew the mission.

By the fourth day the romance was really over. We had grown accustomed to some of the most spectacular sights in the universe. In the beginning, if we weren't tied up on a task, we all tried to view anything unique or spectacular that passed by one of the five windows —the typical tourists. With attitude control fuel in short supply, we spent most of the time drifting and taking pot luck as to when and what would be in view.

On day four, as we were slowly tumbling above the earth, the window on my side of the spacecraft was pointing toward the earth. Wally, trying to decide whether to push his nose against the window also, asked, "What can you see?" Without thinking, I responded casually, "It's just those damn Himalayas." Wally didn't even blink, but a few seconds later we both burst out laughing when we realized how jaded we had become.

By day four we had also become accustomed, as well as we were going to, to our irregular sleep cycle. This was the first spacecraft in which the crew was free-floating while asleep—free-falling is more apt. During all previous flights crew members had gone to sleep strapped in their seats with their hands clasped across their chest to keep from inadvertently flipping switches—a technique I dutifully practiced in the month before launch. We had sleeping stations beneath the left- and right-hand couches. The sleeping restraint functioned as a cocoon which merely prevented the sleeper from floating around the spacecraft while asleep.

The first couple of nights were less than ideal, but I adjusted by simply deciding to quit worrying about being strapped down in order to feel something against my back. The third night was the first that I really slept soundly, for six or seven hours. The sensation of going to sleep and not having a bed—or anything else—under you is one I can best compare to the nightmare of falling through space and never hitting. Just prior to dozing off our subconscious would signal the falling sensation and we would be fully awake with arms flailing. But after adjusting to zero g, it's better than any Posturepedic or feather bed I ever slept on.

In order to keep one man on watch, we never slept three at a time. Wally and I slept simultaneously—he always made a joke of insisting that we didn't sleep "together"—after Donn's sleep period. The three of us would share a watch for a period of eight to twelve hours; then Donn would begin his sleep period. When he woke up, we'd sack out. After our period was up, Donn would wake us and we'd start the cycle again.

Normally, no one slept an entire period, and I was invariably up before it was time. After a couple of problems developed while Donn was on watch, Wally and I slept with one eye open.

Early in the mission, after sixty-one hours, we had a malfunction in the electric power system which threatened a complete electrical shutdown while Donn was on watch and Wally and I were supposed to be sleeping. I was sleeping so near to consciousness beneath my couch that I was able to rouse myself and correct the situation before Donn could get involved. The electric power system was my specialty. A day or two later, Wally and I each awoke at different times to find Donn asleep during his watch. He was floating asleep up in the docking tunnel, in the zero-g fetal position: legs, back, and arms all slightly bent. It was not the best way to be watching the spacecraft. On later missions it became standard practice for all three crewmen to sleep simultaneously, so in retrospect it wasn't critical. But on this mission it didn't make Wally and me feel very secure, knowing the spacecraft was brand-new and unforeseen problems could arise at any moment.

In defense of Donn, he had a more difficult time than Wally and I had, in at least two respects. The only time the spacecraft was completely quiet was during long portions of his watch. Inside a vehicle that small, every little noise is heard throughout. Equally important, the stopping of any usual sound can create an instant alertness. With very little activity—little conversation with the ground, few pictures to be taken, little activity in the cabin—it wasn't too difficult for Wally and me to get a good night's rest. But for Donn, the long quiet periods made it difficult to stay alert and awake. When Donn was supposed to be asleep there were two of us at work and moving about in the spacecraft. The level of activity was reduced, but Wally wasn't the type to accept gracefully too many restraints on what he could do. There were whispered conversations between Wally and me, as well as periodic communications with ground control, all of which made sleep difficult for Donn and caused him to be up and about three or four times during every sleep period. Naturally, that left him still tired during the sleep period for Wally and myself.

Halfway through the flight we had accomplished more than 75 percent of our objectives—the missions were always front-end loaded—and we began to just check off the days rather than exult in them. The flight plan dictated when the day ended. When it said, "sleep," the day was over. But our sleep periods were moved around the clock, from 5:00 A.M. on the first day, to 5:00 P.M. by the tenth day. (We continued to relate our time to the Cape, where we had spent our last several weeks and would return after splashdown.) At the end of a day we would scratch a mark through the date on the calendar. But by October 17, the seventh day, Wally was counting the days so anxiously that he began crossing them off soon after we finished breakfast. We were beginning to anticipate that eleventh—and last—day.

Mealtime was not one of our highlights, just another item on the flight plan. Often we ate on the run, one guy eating while another did the work of the moment, then trading off. We were on a four-day menu, which then repeated itself. By the time we hit the first reruns it was trading time, knocking the hell out of a planned menu that probably cost millions of dollars to develop. But we ignored the diets and kept trading—I swapped my banana pudding for Donn's chocolate—trying to create a little variety and keep our interest in eating alive. This certainly didn't thrill the doctors, but our crew—and most others—looked at eating pragmatically. We treated it primarily as body fuel and only secondarily as a medical experiment.

We had one piece of luck. In the suit room before launch we had each placed in a pocket of our pressure suits several small packets of food, including some freeze-dried, bite-sized bacon squares. This was in case the mission had to be terminated early and we were unable to unpack the food. It came in handy on the ninth day when we discovered that the mission planners had actually slighted us one complete meal apiece, in effect leaving us without breakfast on the eleventh day. We reshuffled the menus so we would end up with the right number of meals, even if they were smaller. The last few meals were looking a bit lean when, on our last full day in orbit, we remembered the packages we had stashed away. We dug those out and our morale leaped when we saw the bacon. Happiness was finding dried bacon squares for breakfast on the eleventh day.

The same mix up that left us one meal short also resulted in the spacecraft being shy one of the cannisters for removing carbon dioxide from the atmosphere.

For the most part we existed on freeze-dried, reconstituted, and bite-sized foods, plus liquid drinks. We didn't have the instant foods that were available on later flights, the kind that were spooned out of an open container. The only hot food we had was what we could mix with the hot water on board, and by the time it was thoroughly dissolved it was no longer hot. It was interesting that after the first couple of days in orbit we began to treat water and bite-size morsels exactly the same. No matter which way we faced, we could just as easily put the water in our mouths before or after the bite-sized dried food. After eating, of course, we could brush our teeth—so long as we swallowed all of the toothpaste afterwards.

The big surprise was finding that the flight plan we had lived with so long had Wally and me eating not breakfast but dinner right after waking up from the fourth day on.

The work-rest cycles made for one giant case of jet lag. The real workday on Apollo 7—where the lion's share of activities were concentrated—began after Wally and I completed our sleep period. This was at 9:00 A.M. on the second day and moved back to 1:00 A.M. on the morning of splashdown. Donn's problem was compounded. In order for all of us to get adequate sleep, Donn had to be hitting the sleep restraint as much as sixteen hours *before* Wally and I were scheduled to get up. That wasn't so bad the first night; Donn went to sleep, or I should say, to bed, around 9:00 P.M. Within a week, when Wally and I were waking up at midnight, Donn was scheduled to hit the sack at 10:00 A.M.—a virtually impossible adjustment, even for Donn, who could usually sleep any place at any time.

By the ninth day Donn was so punchy, and all of our days so screwed up, that Wally and I were able to convince him that it was 10:00 P.M., and that he was overdue to hit the sack. He was headed under the couch before we called the gag and admitted it was really 10:00 A.M.

More than once during my own sleep period I recall looking up with one eye to peek at the instruments for the electric power system and seeing Donn floating above the couch, sound asleep with a pencil in his hand, looking for all the world as if he was working.

After the sixth day it became obvious that we would run short of film before the flight was over and we began limiting our pictures to sights that were, well, meaningful. So what happened? Wally or I would wake up in the middle of our sleep period and there would be Donn, clicking off pictures of the islands out in the middle of the Pacific. One time the movie camera was cranking away as though there was no tomorrow, filming a path across the Pacific, and all we could see down below were clouds.

Our first order of business on waking up was to play Pass the Urine Bag (technically, the UCD, or Urine Collection Device). After it was filled, a series of valves permitted us to dump the contents of the bag overboard.

Now, *there* was something worth taking a picture of. If one dumped just at sunset, the ice flecks coming off the urine dump nozzle would look like a million stars and it would be impossible to take star sightings for about five minutes.

Of course, it's a real experience to see your own urine take on a cosmic quality in space. But it *is* eye-catching and every crew has taken pictures of it. The ice particles are quite beautiful, the same phenomenon that caused John Glenn to rave about the fireflies at sunset. It was considered a mystery at the time of the first reports, and I don't think space officials ever released a clear explanation.

Believe me, just *doing* it was no simple process. A modified condom was attached to one's penis and a storage bag was filled through a valving system. When it was dumped overboard we allowed it to drain for a while to insure all the moisture was cleared out of the system. We were forever forgetting to close the overboard dump valve, and eventually the pressure in the spacecraft would bleed down a small amount and an alarm would go off, indicating an air leak. After several days' use, this alarm came to mean simply that the last one at the john had left the urine dump valve open. When time was short, it was possible to urinate straight through the system, but with a perfect vacuum at one end of the hose and an essential part of one's anatomy on the other, we tended to be a mite careful in manipulating the valves.

Here is probably as good a place as any to answer the other half of the second or third most popular space question. It was the *solid* waste management problem that we all dreaded. Before Apollo most missions were less than four days in duration and it was possible to go on a low residue diet and avoid the problem entirely. On Apollo 7 we made it to the third day before anyone had to answer the call of nature. After my first experience I wrote a memo in the log to help follow on crews preparing for their maiden voyage.

Use of Waste Management Bags

Allow yourself at least forty-five minutes and make as much room as possible in the lower equipment bay. The key to this exercise is setting everything up prior to commencing the main act.

The rear door in your long johns is too small to allow a good placement of the bag. I suggest dropping your drawers and getting comfortable. This means disconnecting the biomed harness. I have done it both ways and the latter is best. Separate the outer bag and stick it to the wall. It will be used to hold all the little pieces of paper that collect. Lay out at least 4 folded paper towels and the wet wipe that comes with the bag. These paper towels will be used, folded, and used again. Wet one of the towels *and* the wet wipe which always seems to be dry. These should be placed within easy reach.

Remove the chemical biocide packet and place in the bottom of the bag. Stick the bag to the appropriate spot on your rear as accurately as possible. Try to keep the one finger glove out of the way. Now attach the urine collection equipment and you're ready to go (pardon the pun).

Use the finger of the bag to push and hold everything at the bottom of the bag. Leave the bag stuck where it is while you clean up the urine portion of the job.

Unless you have been singularly unfortunate you should be able to safely remove the bag from your posterior and attach it by one corner to the couch. Use a couple of the dry towels judiciously and shove them carefully into the bag. Next the wet paper towels and the dry ones again as necessary. Finish with the small wet wipe and close the top of the bag *except* for a small hole. Break the chemical container at the bottom and mix well, trying not to let too much of the chemical get soaked up by the paper.

Knead the bag until well mixed (or as long as you can take it) and roll it tightly, forcing the air out the hole you have left at the top. Place this in the outer bag with all the little pieces that have accumulated.

Remove the inner seal cover, seal it, and roll up the outer bag. Remove the protector on the outer sealer, roll the outer bag and store it in the fecal storage compartment.

There was more than one thing a fecal bag could be used for. Along about our fourth day in orbit I encountered a bag of chocolate pudding that was failing at the seams. In attempting to eat the pudding it was getting squeezed out at several places along the edge. That was an impossible problem in zero gravity and Wally suggested a fecal bag as the perfect solution where it wouldn't smell or possibly explode as it decomposed.

All used fecal bags were sealed, rolled tightly and stored in a special compartment that could be vented to the outside to eliminate disagreeable odors. By the end of the mission we had thirteen little nuggets stored in the compartment and we didn't give them another thought until the day after splashdown. On that morning the three of us climbed aboard an airplane for a two-hour flight from the carrier to Cape Kennedy. The only cargo besides the flight crew on this special mission was one package of air freight. It was a large plastic bag containing those thirteen "nuggets" for early delivery to the doctors back on the beach. Depending on the individual scientist, it was debatable which package the doctors were more interested in. That plastic bag would capture our own attention three days later.

With no shaving for eleven days, our ablutions were followed by dressing for the daily "Wally, Walt, and Donn Show." With only one change of clothes for eleven days, we got dressed and undressed a surprising number of times. For sleeping we always got out of our inflight coveralls and wore our long johns. If possible, we peeled down in the morning to enjoy the luxury of a sponge bath. To answer the call of nature three or four times during the mission there was only one way to go and that was strip all the way down.

We were not exactly in a state of euphoria, but with the exception of those rare moments of tension and concern, our spirits remained high. Every innocent thought or act seemed to inspire some kind of reaction. When Wally held up another of our printed signs to the on-board TV camera—"Keep those cards and letters coming, folks"—it was no time at all before the CapCom radioed a frantic SOS from Wally's secretary. She was being swamped with mail.

Early in the flight, Eisele reported an intriguing experience with a "mysterious" radio signal he was picking up. The transcript of his conversation with ground control went like this:

EISELE: We have some music in the background. Is that you?

GROUND: You must be picking up the twilight zone there.

EISELE: Is someone trying to plug in a radio program to us, or are we just picking that up spiritually?

GROUND: That must be a spurious signal. No, we don't have anything like that.

EISELE: OK, I'm getting a hot tip on some hospital insurance plan from some guy.

GROUND: Maybe they are trying to tell you something.

EISELE: Yeah. Maybe they know something I don't.

And so it went: peaks and valleys, a little mischief and, occasionally, a sense of the universe we were now exploring. Our universe had contracted to the spacecraft and the mission in which we were immersed. Oh, the world went on around us, but it seemed to exist outside the crystal ball in which we were living. The war in Vietnam was a hot election issue ("Don't take any pictures over Hanoi") and Jacqueline Kennedy announced her engagement to Aristotle Onassis.

Seven's response to that news was typically Schirra, "Beware of Greeks bearing gifts."

Throughout the mission we compared notes on what we were experiencing. There was never any clear agreement on whether it seemed like we were flying inverted over the earth or diving down under the planet while flying upright. In fact, at different times, we each experienced a variety of impressions. I came to see the spacecraft and the earth as two unique bodies moving through space on more or less equal terms.

Other subjective effects which we discussed and investigated as time permitted were the "break-off phenomena" and space sickness. In the early sixties psychologists hypothesized that man was psychologically attached to an earth-centered frame of reference. Getting away from this reference, in space, would supposedly create a lost feeling—a melancholic effect. It would supposedly be amplified in circumlunar space. That one was so far out we had trouble even identifying with it. Through the flight, I concluded man existed in a *self*-centered frame of reference. Through his visual sense (primarily) and his brain, he can relate himself to any convenient external system.

Nausea, or space sickness, is a very subjective experience. Donn and I made cautious attempts to induce nausea with no success. Yet half the Apollo crew members had some manifestation of nausea early in their missions.

Finally, Wally marked off the last day, and after wondering in the early stages how the time would pass, it *had*.

As we approached our retro-fire burn, deorbit, and reentry, Wally braced himself for what would be one of his big moments of the mission, second only to the abort responsibility during boost.

An aviator earns his pay by landing the airplane. The degree of success with which he can perform this function, under *any* conditions, is a fair measure of his worth. In flying the Apollo spacecraft the opposite was true. Landing the spacecraft—flying the reentry—is a more-or-less routine, manual-control task, although a certain mystique has developed around it. Furthermore, it can be flown just as well, if not better, by the on-board computer.

On the other hand, many problems can cause an abort during boost, with endless factors affecting the decision. Only two of these problems triggered an *automatic* abort, when an emergency could occur so rapidly that there would be no time for men—either in the air or on the ground—to judge the situation and react. Those were vehicle rotation rates exceeding limits, or two of the eight engines going out. In either case an escape system would blast the spacecraft free and return it to earth.

When to abort and when not to are subjective evaluations that most times are at the total discretion of the spacecraft commander. In this role Wally was virtually flawless. On Gemini 6, he was awarded a medal for *not* aborting during an unsuccessful launch attempt. When the Titan booster belched smoke and flames without lifting off, Wally correctly decided that there was no danger of an explosion. He made a split-second decision not to pull the seat ejection ring when few would have faulted him for doing so. A few days later, Gemini 6, still intact, blasted off, carrying Wally and Tom Stafford to America's first space rendezvous.

Our retro burn was computer controlled, and if it came off properly we would be down in twenty-nine minutes, come what may. The most critical requirement was to be in the proper attitude at the time of the burn. The only activity on board which even resembles flying an airplane is maneuvering the spacecraft. Consequently, we showed a distinct bias for performing maneuvers manually whenever possible. They were planned ahead and performed slowly to conserve fuel. It wasn't unusual to acquire the proper retro attitude well in advance. With that detail taken care of, the crew's mind was freed for other procedures. Besides, it gave one a warm feeling of reassurance just to be there.

When it came time for Wally to get into retro attitude, which would trigger all the events for reentry, he did just as everyone else before him. He started early and got there early, and it was a good thing too. Before we made the retro burn, Apollo 7 had assumed three different retro attitudes. Twice Wally was pointed 180° from where we belonged. And twice Donn and I, by nudging and eye contact and hemming and hawing and mumbling a lot, managed to disconcert Wally until he discovered the problem and finally went to the proper retro attitude.

Those were really the most awkward moments of the flight and potentially quite fatal. We were spared great embarrassment by the fact that we did it early and were out of ground contact for the entire time we were stumbling around. It also gave Donn and me a new appreciation of why one goes to the retro-fire attitude early.

The retro burn itself came off almost as an anticlimax. A few minutes before 6:00 A.M. on October 22, 1968, Apollo 7 made its last pass over Houston, at a height of 76 miles, a small, swift dot of light in the morning darkness. Schirra, staring out a window, reported, "We're flying a pink cloud." And then, his final quip of the flight: "Our landing gear is down and locked."

From the time Apollo 7 started her long slide down the pike, I was already feeling melancholic and cheated. Something I had gloried in and had truly loved was being taken away. It would be a long time, I thought, before I would experience that feeling again. I didn't know at the time that I would never return to that unique world. Before the experience was even over, the nostalgia was sinking in.

Twenty-four minutes after we started reentry, the chutes opened in a solid overcast at 10,500 feet over the Atlantic. Then we were in the water, and eleven days in space suddenly seemed like eleven minutes.

The voyage was already assuming that dreamlike quality of a moment long ago in another place, as though I was remembering scenes from an old movie. In the postflight debriefings it would be easy to describe events or tests or incidents that had filled our time, but impossible to describe the experience itself. It had happened to another Walter Cunningham. Trying to recapture just the wonder-world sensation of zero gravity seemed hopeless.

Eleven days of living together, three men in a superpowered thimble; zero gravity; being for one chunk of time THE news the world over, and a focal point in the lives of people throughout the world—it was already history and beginning to recede fast from my consciousness. The letdown from Olympus had begun.

Splashdown came at 6:24 A.M. We slipped into the water as cleanly as a newly launched ship. We had pulled it off and were pleased. Probably 90 percent of the most critical events of a flight, from the standpoint of crew safety, occur in those last thirty minutes. A safe landing requires retro-fire going off smoothly; satisfactory separation between the Command and Service modules; a properly executed re-entry; ejection of the cone covering the parachutes; deployment of the drogue chutes at 45,000 feet, the pilot chutes at 25,000, and main chutes at 10,000 feet; and finally a good splashdown.

In nominal circumstances there was little the crew could do to improve performance but a lot we could do to screw it up. We were very nominal, and we were down. I piped up with the fighter pilot's wry comment; "Well, we cheated death again."

We might have laughed, except that immediately after we hit, the Command Module began a steady and rapid rotation, flipping us upside down—or at least three-quarters. It left us hanging from the couch straps and looking out the side windows at a very choppy sea. But we had procedures for this and after letting the spacecraft cool for seven minutes, we needed an equal amount of time to inflate the uprighting bags that would return us to an even keel—more than enough time to get us sicker than dogs in spite of the motion sickness pills we had taken just prior to de-orbit.

There was a postlanding checklist we still had to perform, but for now we were just hanging on, waiting for the spacecraft to upright. We also had planned on removing our pressure suits and slipping into our cleanest set of coveralls. After hanging inverted for three or four minutes, with each man keeping his own counsel with his stomach, one is not only hesitant to move but even to speak. Then Donn asked, "How do you feel? You going to get sick?" Without waiting for an answer, he added, "I think I'm going to throw up." From that moment on I really had squirrels dancing inside my stomach. Even Wally was a bit concerned, and he had cast-iron guts.

Donn, bless his soul, had the most difficult job of all. As soon as possible after landing he was to get out of the couch, go down to the lower equipment bay, and connect a VHF radio antenna. It was close quarters, head down in the bottom of the craft, and no circulation. Donn would invariably get sick before the job was through and it would fall to me to complete it; and within a few minutes the chances were good that I'd be as sick as Donn. But it was all academic. Donn got sick just hanging in the straps, while Wally and I fought off nausea until we finally got the hatch open.

There was a brief spell of apprehension when we noticed water pooling in the lower end of the spacecraft, and seeping between the inner plates and the exterior windows around the shell. We didn't know if the water had leaked in through a ruptured pressure vessel or through a ventilation fan, or if it was simply condensed moisture. We kept an eye on it, until each of us stepped happily over the side into the frogmen's raft. The Command Module was a great spacecraft, but on the water it was more like a cork than a boat.

In the chopper we began to realize how fortunate we were that the reentry had been accurate and they were standing by at our splashdown. The weather was 600 feet overcast, with two miles of visibility in rain showers. It was no problem, but it had cheated us out of looking up and seeing the gorgeous sight of our three main orange-and-white parachutes drifting down against a background of blue sky.

Once we stepped onto the deck of the carrier *Essex*, I began to shake off the postorbital remorse (which must be akin to postnatal depression). After eleven days in space, when you see the U.S. Navy turned out on a carrier deck, it strikes home that millions of other souls have been sweating out your success and your safety. My feeling at that moment was, "Thank heaven we came down in our prime recovery area." It would have been unthinkable to have disappointed those people by landing in the Pacific Ocean.

Then there were the cards, letters, and telegrams waiting for us on board. They dramatized how many people had shared our flight as a personal ordeal—and triumph. It made me wonder what I could possibly have done to warrant such support, how to justify all the faith invested in me.

Physically, the one overpowering sensation we experienced was one of heaviness. When the postflight medical checkup started and I lay down on the examining table, it felt as if I were going to sink right through it. I felt as if I were made out of lead, and this feeling persisted until the next day.

As I lay there, I looked up at the overhead, which was several feet above. It seemed strange that I couldn't do what I had done for eleven days: push off from the table by pushing with one finger or flexing my shoulder muscles and simply float to the ceiling. I imagined myself sitting there, upside down, and holding to the pipes to keep from floating away. No more, not anytime, no how!

Even our clothes seemed heavy. When we stepped out of the helicopter onto the carrier, we all had the uneasy feeling that our pants were going to fall off. I hitched mine up and the three of us grinned crazily as we realized why.

I began to feel weary: not that spaghetti kind of weary, where you just want to melt to the ground, but the kind where all your bones send you a message of a physical job well done. The scuttlebutt around the office had been that an early prime crew might be recycled for the first lunar landing. But I had no illusions at that moment and I knew it wouldn't be us. It was virtually impossible for any crew to fly two of the big missions. I made it a point never to kid myself. But one thing is certain: it is possible to experience a decade of manned space flight, as has happened to some of the men, without leaving a mark on it. The first manned test of the Apollo spacecraft offered an opportunity to be counted and who has a right to ask for anything more? I'd rather die than not be counted.

A near-perfect mission was behind us. Suddenly the goal of landing men on the surface of the moon was very close. After a two-year hiatus marked by tragedy and then triumph, Donn and I were Astronauts. For the next sixty days, until Apollo 8 lifted off, we would be the voices of authority. We had been there and joined the "Brotherhood of the Right Stuff." No longer would we go out to show training movies and talk about our buddies' space flights. From now on the reports would be first hand.

Riding the Hero Trail

THE WRITER Tom Wolfe once referred to it as postorbital remorse. The astronaut most often cited as the classic example is Buzz Aldrin, who, in the opinion of his admirers, including himself, was unable to cope with the hero's role and the adulation that followed the moon landing of Apollo 11.

To all of which I say, "Hogwash!" Buzz signed into a hospital when he found himself heading for what he labeled "a good, old-fashioned American nervous breakdown." It took nerve to do, and even more to make it public. But I don't believe the "incessant round of appearances," the glory, or the applause were Aldrin's problem any more than they were for the rest of us. The problem comes when one begins to hear his own kind of music, likes it, and then realizes it won't last forever.

It is easy to get the impression from the story of Buzz Aldrin that all the astros were vulnerable to a kind of postflight depression. Let me set the record straight.

If there was a letdown it was professional, not personal. We had completed the mission, a goal for which we had trained and dedicated and sacrificed most of our lives, but there was no way to know if we'd have another. And besides, part of our training, whether intentional or not, included traveling and making public appearances. If the social fanfare and public obligations were the sources of the problems some had, we very possibly wouldn't have lasted through the program.

Early in the Apollo program it became standard practice after each mission for the crews to take off on a grand tour covering key cities here and abroad. The obvious aim of this dog-and-pony show was to spread good will for the space effort and the United States. NASA hoped to get the maximum value from these excursions, and their imagination was limited only by the amount of cooperation they could expect from the flight crew. In a sense, then, the astronauts themselves could influence the scope of the postflight activities and they were in a fine position to bargain: they'd show a little more enthusiasm for NASA's desires in return for the approval of an appearance or a speech that would help a friend, or a business contact.

The real prize in the postflight reward league was the grand tour. These were not just sun-and-fun-filled vacations out of some travel folder; most of the days were long and tedious and filled with official functions. But I don't know a crewman, or a crewman's family, who wasn't anxious to make one.

How elaborate a tour it was depended on NASA policy at the time, the national and international interest generated by the flight, and just plain timing. Certainly the Apollo 11 crew, the first lunar expedition, was in more demand than any group or individual since Lindbergh. Their tour represented a masterful job of planning by NASA and the State Department. Phileas T. Fogg would have wept with envy. And I suspect Buzz Aldrin wasn't as reluctant a passenger as he made out. The tour is described ad infinitum in Buzz's book, but it can be summarized as twenty-three countries in forty-five days—with a footnote for what Joan Aldrin categorized as "three kings in two days."

That was heavy company and the schedule was hectic. But other crews—notably Apollo 13, whose narrow escape from misadventure captivated the public—endured schedules equally overloaded. The Apollo 11 crew did carry the brunt in 1969, but in the years immediately following, the number of flights was cut in half while astronaut appearances nearly doubled.

It was briefly a source of mystery to me that the flight of Apollo 7 inspired so little postflight activity after such tremendous preflight interest. The exchanges between Wally and Mission Control had turned off more than a few people but I also knew that NASA had to keep the bandwagon rolling. Requests for the crew were being fielded, as usual, by the public affairs office at NASA headquarters. As rookies, Donn Eisele and I were not too proud to venture out among the natives and receive their homage. Maybe a little parade, or something. Wally, on the other hand, after three missions, had ridden in enough parades to last him a lifetime. Anything short of a New York tickertape parade, with the Radio City Rockettes in his lap, would probably not have been acceptable. The one parade being pushed by NASA was in San Francisco, California, which happened to be my home state.

Without our knowing it, Wally had left orders at NASA headquarters that all postflight activities were to be channeled through him. When the pitch was made for the San Francisco parade, Wally quickly vetoed it. Poor cooperation wasn't going to improve a reputation earned in flight as NASA's tarnished heroes and we eventually coasted through one of the least demanding postflight calendars of any Apollo crew.

The closest Lo and I came to one of the "grand tours" was a ten-day swing through Australia and New Zealand in the first part of May 1969. As a kind of minijunket, it was representative of the treatment the astronauts received abroad, and the itinerary was a challenge. Lo and I number it among the more memorable experiences of our eight years in the space business. The occasion was the annual celebration of the Victory of the Battle of Coral Sea, in which American and Australian forces fought side by side in a naval engagement that became a turning point in the Pacific war. It began May 4, 1942, ended four days later, and saved Australia from invasion by a Japanese armada.

We were there in a unique social-diplomatic role. It had been traditional for high-ranking military officers to represent the United States at the anniversary celebrations that took place throughout the Australian continent. That year, with tension rising over certain political issues, including the war in Vietnam, and in an effort to play down the anti-Japanese implications of previous ceremonies, the Australian government had indicated that a civilian representative would be preferred. An astronaut made the likely candidate, which is where I came in.

Other departures from the past promised to make the trip a trying one. In earlier years the dignitary of the moment—an admiral or four-star general—would tour the seven largest Australian cities over a period of three weeks, reinforced by vessels of the Pacific fleet or the air force band, with appropriate regalia. In 1969 it was compressed to five cities in seven days; there would be no Pacific fleet or air force band, but only what we could muster in the way of support from the U.S. Embassy. In effect, it would be *a cappella*. Then there would be a number of school appearances, added in deference to the intense interest of youngsters in space exploration. And, finally, Lo and I would be traveling without the usual entourage of protocol people.

One incident during the trip gave us a fair insight into the entire celebrity syndrome. There seems to be a kind of orbit in which celebrities move. When one has reached a certain level of notoriety, for whatever reason, he becomes handshake material. Total strangers

know who you are, or think they do, and assume that a sort of fellowship exists among so-called public figures—that they know each other,
or would like to if they don't.

We arrived at the Chevron Hotel in Sydney one night to find the
lobby swarming with people. The manager explained, "Tom Jones is
opening tonight in the lounge."

I brightly asked, "Who's Tom Jones?" The name meant nothing to
me.

The first show was due to go on in thirty minutes and when someone
asked if we'd like to see it, we jumped at it. They were filling that
room, six hundred seats, three times a night, standing room only. Now
someone had half an hour in which to find a table for six, front row
center.

In the meantime, the manager insisted we meet Tom Jones, assuming, of course, that it was a courtesy to us both. While we were unpack-
ing, Jones was brought from his room, past the throng of teenyboppers
that camped in the hall of the penthouse floor, and to our door. I knew
nothing about Tom Jones, and he obviously had never heard of Walt
Cunningham. We shook hands and tried to make chitchat. We really
had nothing to say to each other; the one thing that passed between us
that had any sincerity and meaning was our mutual appreciation of the
awkwardness of the ritual. We both knew that each had been dragged
through that same routine dozens, even hundreds of times, and no
doubt would again.

My sympathetic neutrality turned quickly to admiration when he
proved to be a superb showman.

The eagerness of the hotel manager was a common affliction. The
public is familiar with the attraction and popularity of an entertainer
and they identify with that part of an astronaut's life as well or think
they can. With the entertainer, people see not only the personal acclaim
but its reason for being, the public performance. It's the same with an
athlete: an O. J. Simpson, or a Mark Spitz, or a Johnny Bench.

With an astronaut it's not so simple. The acclaim is the same, but the
only direct exposure the public gets is the few hours in an astro's life
when he is "on stage" during his mission. And even then, what he is
doing isn't clear to most, let alone what he does with the other 99
percent of his time. The astronaut mystique more closely resembles
that of a well-known politician than that of an entertainment or sports
figure.

And that's the point. The ego trip provided by the worldwide attention lavished on those returning from space can be intoxicating stuff.
It's easy to get high on the flattery and flourishes and the self-satisfac-
tion of a job that is unique, no one is totally immune. The trouble

comes when one forgets that the acclaim is really for the event, and not the individual. When the applause dies down, as it always does, the withdrawal symptoms begin.

Throughout our visit Down Under, we traveled in style in the personal jet of Prime Minister John Gorton. His BAC-111 became the flying command post for our traveling party, which varied between six and a dozen people. The significance of being the official United States representative in Australia was brought home when the captain of the aircraft solicited my permission in Melbourne to carry the minister of supply—the equivalent of a cabinet post—to Canberra, our next stop.

Wherever we went, the schedule was laid out with stopwatch precision. The first stop, Brisbane, offered a typical agenda: tea and biscuits with city officials and officers of the Australian-American Association; a press conference; a parade; the laying of a wreath to honor the dead of World War II; a luncheon with speech and receiving line; a visit to the local racetrack for two races and an introduction; an appearance at a rugby match with a parade around the field at halftime; a presentation to a combined assembly of local high school students and talk show audience; back to the hotel to tidy up and dress for a black-tie dinner-dance and speech; then to bed around midnight to be up the next morning to start all over again in another city.

The Australian people were open and friendly, very much like in the American Southwest. From the beginning there was a strong identification with the American people and the space effort. The first person to greet us on arrival in Brisbane was the Right Honourable Lord Mayor, Alderman Clem Jones. His first words to me were, "So you went to UCLA. Well, I graduated from there myself."

That proved to be typical of the Aussies in their attitudes toward Americans. I encountered nothing but friendliness, even though the Australian government at the time was embroiled in problems surrounding the use of U.S. Air Force tracking sites in the interior and a controversy over the purchase of American fighter planes.

I did finally get my very own "hero's" parade through downtown Adelaide, and it isn't one I'll quickly forget. To begin with, Australians are a rather blasé folk not given much to pomp or finery or frills. Their idea of a good time is to drink their share of beer and watch a little "footy" (Australian rules football). Lo and I were seated in the back of an open Rolls-Royce that had been brought down expressly for the transportation of Queen Elizabeth on one of her visits. Our instructions were to stand up and wave at the spectators, but it was immediately obvious that the spectators didn't give a toot whether we were there or not. It was a Saturday, and they were clearly intent on their shopping chores. We drove through town trying to attract attention by waving and smiling and otherwise looking silly while the good folks of Adelaide went blissfully about their business. The Australians must have felt that a touring astronaut was entitled to a parade, and it was nice to be thought of as parade material. The awakening was in finding out that I wasn't.

Parades, of course, are for the little kid in all of us. But at times they can get hilarious. The Apollo 12 crew came back from their world tour telling about a parade they had in Iran where the local Communist party managed to put out some false information on the parade route. Pete Conrad, Al Bean, and Dick Gordon found themselves headed down the parade route while the crowds were lined up one block away wondering where the astronauts were. In some of the countries that were visited many of the people don't read. They may know nothing of world events and haven't the foggiest idea why those idiots are riding along in those great huge cars.

From Australia we flew to New Zealand for three more days of celebrations and a civics lesson. New Zealand is probably the most socialistic country in the world, and plagued by the frequent strikes that are characteristic of that form of government. There were several going on during our brief visit. The one that attracted my attention involved the pilots of Air New Zealand. The salaries of the captains on Qantas, the Australian airline, had recently been raised to $23,000 per year. The captains on Air New Zealand, by contrast, were paid $12,000 a year to fly the same equipment over the same routes. They felt justified in striking for equal pay.

At a luncheon our first day in New Zealand I sat next to Prime Minister Keith Holyoake, to whom I had presented a flag that had traveled with me on the five-million-mile trip of Apollo 7. While making lunch talk, I happened to mention—tactlessly, I suppose—how surprised I was to learn that captains on Air New Zealand earned only $12,000 a year. He replied, "That places them among the highest-paid people in New Zealand." Then he went on to ask me, "How much do you get paid in a year, Mr. Cunningham?"

That caught me off guard, but I answered, "Oh, eighteen to nineteen thousand dollars." The prime minister then informed me that was some $6,000 more per year than he was paid. He received the equivalent of $12,500 a year in salary with an additional household and entertainment allowance of about $6,000. The "first family" lived in their own home, run by his wife, with no outside help. He made his point. From his perspective, the airline pilots were asking for $5,000 more a year than the combined salary and allowances for the chief executive of the country.

In one respect, the trip led me to take stock of the isolation booth the space program had become for me. The month that Lo and I were together, including the few days before and after the trip, represented the longest continuous period of time we had spent in each other's company in six years. I had felt all along that our marriage was healthier than most and well able to stand the assorted pressures. Now I became conscious of how wide the distance between us had grown in terms of my job and what had been happening to me, and how much Lo had been shut out of it. We had damn near gotten out of the habit of communicating. I made a note to myself to back off and get the right perspective on the things that counted, starting with my family.

There were times during our Australian swing that Lo would catch me studying her, and I'd flash her a wink. I was thinking of how she had made things so much easier for me in so many ways. The weeks immediately before a flight are loaded with potential stress for an astronaut's family. How the wife reacts can either ease the pressure or compound it. Lo came through beautifully. She was lighthearted as the days passed and on the morning of splashdown her first comment to reporters was, "Walt's just been away for a couple of weeks and now he's back."

During a mission the wives of the crew members are the objects of continuous interest by the media. Often a dozen or so reporters and other media representatives would be camped on the front doorstep. They were interested in getting pictures of the family, anything: the wife, kids, goldfish, parakeet. A brief chat with the lady of the house made their day. This sometimes led to some embarrassing moments for some of the families. Foot-in-mouth disease was not unusual among a group of women who had never been given any briefing on the handling of a press conference. During the mission, however, NASA did provide a protocol officer to coordinate their contacts with (and protect them from) the media.

On some occasions the wife's behavior during a mission would catch some criticism—if not from her loving husband, then from the other astronauts. It was with some trepidation that a crewman caught up on the newspapers that had collected in his absence. He might find himself thinking, "My God, did she really say that?"

I was relieved to read the papers and discover how skillfully Lo had handled herself. She had acted with composure and dignity, had kept the kids under control, and in all the photographs came across as the beautiful woman she is. I marveled at her poise and was proud of her. The flight consumed us all, of course, but Lo avoided the impression some wives gave that the fate of civilization was at stake. She treated it all with just the right touch. It occurred to me how great an adjustment she had made over the years. We both came from poor backgrounds, with no exposure to the world of fame and high society. Today Lo can hold her own anywhere.

The only surprise I received in scanning those back newspapers was the account of a press conference conducted by Gen. Sam Phillips, the headquarters chief of the Apollo program, immediately after we had splashed down. The general was the first to hail the mission of Apollo 7 as "a 101 percent success." The media picked up the phrase, and the headlines and stories over the next several days spread the word.

My first reaction was that the remark had been taken out of context during the press conference. If not, it indicated that the mission was being appraised, for public consumption at least, on the basis of our technical accomplishments and not the angry exchanges between Wally and Mission Control. Yet I could not honestly reconcile that judgment with my own. After spending the last three of the eleven days doing practically nothing except existing, I could hardly conjure up a picture of a performance above and beyond our objectives.

This probably requires an explanation. After all, you can't understand the score if you don't know how the game is played.

The success of a mission is measured against a comprehensive list of mission objectives. Flight crews are aware of this and are particularly sensitive if they have flown earlier missions that didn't yield the magic number of 100 percent success.

In the early days when nothing was known of man's capabilities in zero gravity, and NASA was not yet numb to the cost of doing business in space, every effort was made to have an action-packed mission. The voice of moderation in workload planning has always been the astronauts'. It didn't take us long to realize the surest way to guarantee success was to insure before launch that the objectives were neither too many nor too difficult.

Our competitive motivation to always look good was consistent with NASA's image-building efforts, and a trend was established that would sometimes produce mediocre workloads and high success measurements. There were both good and bad reasons for keeping certain experiments and objectives off the flight. Without extremely close scrutiny by the flight crew, many half-baked and useless ideas would have been added to a mission. There are many proponents of collecting information—any information—just because space is a unique environment. Some information is sought long before anyone has conceived of a way to make use of it.

Such tests, as well as those with a useful but perhaps subtle application, were frequently dismissed by flight crews as gee-whiz experiments. Looking ahead to a triumphant return from the "perfect mission," it was easy to become zealous in eliminating useless, unclear, difficult, medical, or scientific objectives. The best way to avoid failure on a complicated test was to eliminate it from the mission.

Astronauts are generally thought of as adventurers, gamblers, and risk-takers; quite the opposite is true. Our philosophy was to reduce the risk in all aspects of any operation. Not taking on questionable tasks was consistent with that philosophy. Never mind that that attitude worked directly against some worthwhile experiments and engineering tests such as those with complex protocols and above-average chances of being screwed up; experiments whose past efforts may have contributed to embarrassing the flight crew; experiments in which the crew served as guinea pigs; medical experiments; scientific experiments, depending on the prejudices of the particular crewmen involved.

The most diligent efforts of some crews came in the sensitive area of preflight planning. It kept out the more wild-eyed ideas and insured flights where mission success was guaranteed before lift-off. The price for such success was that many valid experiments were dumped and many days during the latter part of a mission were spent with little to do except count the hours to splashdown. But the system produced an unparalleled string of successes.

This attitude gained strong momentum during the early days of Apollo. The crew of Gus Grissom, Ed White, and Roger Chaffee had their own order of priorities for what was to have been the first Apollo mission. They went at it as an engineering test flight and, applying that definition, eliminated anything that even sounded scientific. Since they hoped up to the very last to undertake a mission that would break the fourteen-day space endurance record, they looked with disfavor on anything that might compromise that personal objective.

When our crew was on Apollo 2, we fell heir to everything they had dumped. With Wally Schirra even less enthusiastic about science and scientists than Gus Grissom, his interest sank with each new blivet we inherited. After the fire investigation and the Apollo program restart, we had enough stroke to pull a rerun of the Grissom "engineering test flight" bit all over again. Minimizing the chance of failure were the criteria applied to eliminate those objectives most demanding of time or talent.

The major effort in flight planning is blocking out the times for different activities. Failing to complete an experiment in flight is a major embarrassment, especially when everyone and his brother can perform it in the 1-g environment on the ground in much less time. Things do take longer in zero gravity, but the contingency times we built in could sure add up when the experiments ran smoothly.

Ground planners get their licks in another way as well. The actual duration of any mission is really an unknown and many things could cause it to be terminated early. To cope with this possibility, not only are more important tests scheduled earlier, but also the entire workload is shifted toward the first part of the mission. That is particularly true (and justified) in the initial flight of any new spacecraft.

So while General Phillips and others described Apollo 7 as a 101 percent success, I personally felt we lifted off with a workload planned to about 60 percent of our capability. We would have accomplished a great deal more had we scheduled twice as much activity and completed only 80 percent of it. Some equipment acceptable to us on the canceled Apollo 2 mission (such as the scientific airlock) were rejected for Apollo 7 and didn't fly until the Skylab missions five years later.

Not all crews were alike in their approach to flight planning and scheduled activities. They ranged from a very hard-nosed attitude by the Mercury astronauts ("how to go into space without risking public criticism") to the last group of scientist-astronauts ("stick it to me"). Those late arrivals were unaware of their strength vis-à-vis management, were not prima donnas, were responsive to a broader spectrum of requirements, less aware of in-flight problems, and more willing to be judged on how much, rather than what percentage, was accomplished.

The debriefing period encompasses a minimum of ten days of hectic activity. It begins immediately after the shipboard reception, a phone call from the president, a physical examination, and a shower that can only be appreciated by those who have bathed for two weeks in less than one glass of water. We started our first debriefing tapes before dinner, aboard the recovery ship. These tapes were completed and transcribed and became part of the postflight report.

Verbal debriefings, in some ways, are comparable to a suspect's position in a sordid criminal investigation. They come at you in relays. The first was an all-day show-and-tell with the top brass: program managers, center directors, headquarters people, and the prime contractors. This was followed by special interviews with the engineers and those involved with specific mission experiments, and another day-long session with the other astronauts. In that one we let our hair down, talking and laughing about everything from individual idiosyncrasies on the urine collection device to the things that went wrong which we don't want anyone else to hear about. By the time the media gets hold of a crew, its act has been pretty well polished and the world gets very few surprises.

The serious work of debriefing was interrupted by one humorous incident. While engineers were analyzing systems, Rita Rapp, the nutritionist, was analyzing each of the empty food bags to determine exactly how much each man had consumed on the flight. That was going to be a little tough since a fair amount of horse trading was taking place after the fourth day to help maintain our interest in eating and minimize the weight loss.

Rita was a shy little woman in her thirties, very prim and proper. After three days of weighing and measuring, she finally gave up and called us pleading, "Okay, I give up, where is that last food bag I'm missing?" None of us knew what she was talking about until it was identified as chocolate pudding. The one leaking package, wrapped and stored in a fecal bag, could be pictured now lost among the thirteen others which had been used as originally intended. When we told Rita the bad news that it could be found with the boys analyzing the other end of the food chain, she gave a small groan.

We didn't hear for several days, then finally Rita called to say she had located the lost pudding in the last of the thirteen bags which she had carefully unwrapped in pursuit of her data. If that isn't dedication I don't know what is.

As the debriefings continued, we worked on our postflight pilot's report, one of two dozen sections in the official NASA Mission Report. This document receives a wide distribution, which probably accounts for the fact that NASA likes to keep it pretty well laundered and the content consistent with the space agency's desired image.

Consequently, NASA edited out of our crew comments anything that might reflect negatively on planning, hardware design, or poor performance by any personnel. Not even self-criticism was acceptable. My primary responsibility was that of all the subsystems except the guidance and navigation system, and I addressed myself to those areas in the pilot's report, writing that "accurate, on board gauging of the maneuvering thruster propellant system should be a *requirement* for future space vehicles." *Requirement* was changed to "great asset" in the final report. When I wrote that part of the cooling system (the evaporator) was overserviced with water—my goof—and thus not available for use during reentry, the self-criticism was deleted in the published version.

When reference was made to a mission rule that had been knowingly violated during the flight (the recharging of reentry batteries which resulted in a potentially dangerous condition when we separated for reentry), the final report deleted the entire paragraph. The omnipotence of the astronaut and the infallibility of NASA had to be maintained.

Another part of the postflight ritual was less sensitive. After the flight it was routine for each crew member to be decorated for contributions to manned space flight with either the Distinguished Service Medal (the highest) or the Exceptional Service Medal (second highest). A particularly noteworthy or unique contribution would qualify for the

DSM. When the Gemini 6 crew experienced a power kill on the pad, Wally Schirra's decision not to abort—thus making it possible to re-schedule the mission—qualified him for the Distinguished Service Medal. The time, place, and person making the presentation was determined by NASA's attitude of the moment, the media coverage available, and the political situation. When it was all sorted out, the Apollo 7 crew was to receive its medals from President Johnson at the LBJ Ranch during the first week of November.

I was pleased at the prospect not only for the honor but also for a basic, old-fashioned reason. I was excited that Lo, Brian, and Kimberly —as well as my parents, who were flying in from Los Angeles for the event—were going to meet the President of the United States. On the appointed day the crew and families of Apollo 7 climbed into a NASA Gulfstream and headed for the Texas hill country. President and Lady Bird Johnson greeted us at the jet strip which had been built on the ranch. Our crew piled into a Lincoln Continental, with LBJ himself as chauffeur, for a quick tour of the ranch. Our wives went ahead in another Lincoln with Lady Bird, who chatted with the President by radio.

Later we toured the small ranch house where the Johnsons lived. The place was immaculate and filled with a quiet Texas charm. It was the first opportunity that Lo and I would have to enjoy the warmth and hospitality of Lady Bird Johnson. We were to enjoy the pleasure of her company on several occasions over the next few years, and each time we were impressed with the ability of this very busy woman to make her guests relax and give the impression that she had nothing else to do except make them feel comfortable and at ease.

The ceremony went off smoothly, with awards going to the Apollo 7 crew and to others, including Tom Morrow of Chrysler, the man re-sponsible for producing the booster we used. Wally made a brief speech and presented the president with a picture of central Texas, where the LBJ Ranch is located. I had taken the photograph on our second revolution, but unfortunately the ranch itself was completely hidden behind the Saturn booster in the foreground. After the cere-mony, Wally, Donn, and I went with the president to leave our hand-prints in some concrete stepping stones freshly poured for that purpose. Many such squares, inlaid with visitors' handprints, were used as walkways around the ranch house.

When it was time to leave we had to round up Brian and Kimberly, who had been exploring the ranch with the Eisele kids. As we all walked back to the airstrip, Lo and I shared a typical parental concern. We knew it had been fun for them, but we wondered if Kimberly and Brian had any understanding of meeting the president and visiting his home. It was important to us that the significance of the visit not escape them.

We need not have worried. At the plane, the president was saying good-bye and shaking everyone's hand. When Brian's turn came, he reached up, took the president's big hand in his, and said, spontaneously, "Thank you for inviting us, Mr. President."

Oh, yes, the medals. We were each presented the Exceptional Service Medal—NASA hero, second class. We were proud to accept them and they are cherished to this day. At the time, Donn and I discussed what effect Wally's distemper in orbit might have had on the choice. There was no way of being sure, but later events left little doubt that it was indeed a factor. Every other Apollo crew, ten in all, as well as the nine people who flew in Skylab, received the Distinguished Service Medal.

Because I had been so thoroughly embarrassed by Wally's behavior several times during the mission, one of the first things I did on return was to call those people I had worked with and admired at Mission Control. Personally and for the entire crew, I apologized to John Aaron, the systems engineer who had worked with us in developing the malfunction procedures; Glynn Lunney, the best flight director in the business; and others. Without exception, they assured me that they knew where the problem had been and seemed genuinely glad I shared their sensitivities to the situation. If the working troops who had taken the brunt of Wally's tirades knew and understood the situation, I assumed that all those up the line must understand also. My first doubts as to this conclusion came when I realized some of the guys in our own office weren't too sure whether it was all Wally's show or Wally, Walt, and Donn.

It was months after the flight when I first learned of a rumor that Chris Kraft (at that time head of flight operations) had been overheard to say—during one of the mission's acrimonious exchanges—that he would see to it that none of our crew ever flew again. It was a difficult story to track down, and not one that Chris would have been happy to know was being passed around. At the time, my job as chief of the Skylab branch of the Astronaut Office had me working on a program that was not scheduled to fly until after the lunar landings were over. It didn't promise to be one of the better assignments. On the plus side, Tom Stafford, the chief of the Astronaut Office, had given me verbal assurance that my efforts would assure me command of the first Skylab mission. With that in the offing, Kraft's animosity could be a fatal obstacle. Very few doubted he carried that kind of influence. After all, who could forget Scott Carpenter's short career?

In the spring of 1969 I decided to confront the rumor head-on and met with Kraft. I laid my cards on the table, told him what concerned me, and asked if there was any truth in what I had heard. His reaction wasn't exactly outraged innocence, but he did deny making the remark.

Valid rumor or not, I thought about it when Wally resigned (shortly after the mission), when Donn Eisele was eased out after backing up Apollo 10, and when I resigned in 1971—all of us without flying again. I still believe the entire Apollo 7 crew was tarred and feathered through the actions of Wally Schirra during those eleven days in October. We were never collectively hauled on the carpet, and, according to Lunney, no one ever really took Wally to task either. The feeling was, well, it's over and it probably will never happen again. The only ass-chewing I'm personally aware of is the one delivered by Deke five minutes after we had returned to the crew quarters at the Cape from the recovery carrier. Donn and I weren't invited but we had the impression Wally did a whole lot of listening.

The motivation, attitudes, and behavior of two-thirds of that Apollo 7 crew were different from Wally's. If two of us felt so differently why couldn't the atmosphere established during the mission have been changed? It is necessary to understand the command structure essential to such complex and demanding operations, the military upbringing of the participants, and Wally's personality to appreciate the difficulty in effecting such a change. The roots of the problem and NASA's willingness to accept the situation can be traced to the prima donna attitude that grew up around the Mercury astronauts. Wally was in charge of the spacecraft and he attempted to expand that authority over the entire mission.

It might have been possible to effect a change in attitude during those eleven days. On the other hand, Donn wasn't particularly burned up about the reputation Apollo 7 was generating and I could well have ended up alone in the effort. A full-fledged effort to change Wally's attitude, if unsuccessful, could have created an atmosphere on board that would have been very detrimental to the mission. It could even have led to early termination. When three people are confined in a 225-cubic-foot cabin for eleven days, nothing seems quite so essential as an attitude of goodwill and cooperation. As the situation actually developed it was extremely embarrassing to me, but my low key and persistent efforts to keep it balanced did not disrupt the harmony. We were able to accomplish more than originally planned, although at a cost I did not like. Where professionalism is the name of the game, the entire crew was labeled as crybabies.

It may sound trite, but the mission came first.

The fame we had won as a group really belonged to the job, rather than to the individual. And the ways in which the public confuses the individual astronauts helps to keep us humble. The public has no idea which one you are, and they may not know if you ever flew in space. If you had a flight, they won't know which mission you were on; if they do know, they won't know or care what your assignment was; they won't know if the mission was in earth orbit, or went to the moon, or landed on it. Yet each of us is generally referred to as "a man who walked on the moon." They usually feel awkward at not recalling the mission, but even I can't always remember who flew what.

I have been introduced dozens of times as the man who walked on the moon, although I have been only 300 miles closer than the reader. After a while we cease to correct this misconception and simply say, "Thank you." Once in a while we give in to our feelings. In 1973 I was in Los Angeles giving a reading of "Madrigals for Space" in a concert performance of the Roger Wagner Chorale. On the sidewalk outside the theater, as we hurried to the performance, one of the musicians in the orchestra caught up to me and asked, "Are you the astronaut?" To which I answered, "Yes." His immediate response was, "Which one?"

I couldn't resist replying: "This one."

It is part of a familiar phenomenon. We are recognized, but not identified. I am frequently mistaken for Gene Cernan or Scott Carpenter, both of whom I faintly resemble, and sometimes for Alan Shepard, which puzzles me. I don't have Al's steely-eyed stare, so maybe it's the reputation we share for being blunt and often brusque.

Sitting one morning with Louie Welch, then the mayor of Houston, in the coffee shop of the Houston airport, I was asked twice within five minutes for my autograph—once as Scott Carpenter and once as Alan Shepard.

It would be easy to get caught up in the American way of fame. But the postflight glamor is temporary, the fallout, not the main event. We were standard bearers during the lull between flights. It was a little like being the annual winner of a perpetual trophy. The winner's name goes on the silver punch bowl and possession is retained for a short while, but when next year's winner emerges one's accomplishments are returned to perspective as a line of type alongside last year's date.

If these things are forgotten while the spotlight is swinging to the next crew, especially if one begins to feel that he is the anointed one, the letdown can be traumatic. For some it isn't too much too soon, but rather a case of too little for too short a time.

That's why I can say NASA's postflight schedule wasn't to blame for Buzz Aldrin's breakdown. In the opinion of his peers, Buzz was more concerned about having to get off stage than he was at being on. In that hectic Apollo 11 postflight period, there was a lot of exploitation going on, but there was also a lot of capitalizing. When an option was presented, Buzz volunteered for more appearances than most. One of the girls who worked in the office described it as a "my public deserves to see me" attitude. When it began to wind down, he may have felt someone was stealing his thunder.

I believe the biggest single cause of Buzz's problem was even more fundamental. He was the only member of the Apollo 11 crew who did not feel extremely fortunate at their unique role in history. From the time they finished postflight quarantine, Buzz always seemed to be doing a slow burn. He wasn't pleased to be the second man on the moon; he was annoyed as hell that he wasn't first.

Looking back, I suppose it would have served the mental health of all of us better if we had opened an occasional window and let in more of the outside world. Once the echoes of Apollo 7 had faded, I realized I had missed the entire 1968 Olympic Games, held that year in Mexico City. They never even existed for me, and yet as a sports fan I had always followed the day-to-day results hungrily. But with the hectic days before launch, and our days in space coinciding with the games, they just disappeared into the woodwork. My only recollection of the entire spectacle was one message from ground control, reporting that there had been a twenty-nine-foot broad jump by Bob Beamon and an eighteen-foot pole vault by Bob Seagren. Come to think of it, Apollo 7 probably stole a few headlines from the Olympics that year.

10

There's One Born Every Minute

WE WERE ON a magic sleigh ride, soaring above the rooftops and across the heavens, but, unlike Santa, we were taking on baggage, not delivering it.

In the glory years of space, what the country kept forgetting was that we were people. Each of us was, in fact, four people: adventurer, social lion, would-be business tycoon, and political object.

The one thing we were not was heroes. We didn't look in our mirrors each morning and see the Little Shepherd of Kingdom Come. That was something the public craved and the media had created—not that we didn't enjoy the myth, we just never fully understood it.

In whatever we did there existed that one universal thread: competition. If we took up handball to keep in condition, friendly matches became bloody tournaments. We hustled to see who could plunge faster and deeper into the more complicated business deals. If we had to take a urine test, we competed to see who could fill the biggest bottle. (Once, when a nurse asked Schirra to leave a urine sample, he obtained a five-gallon jug, filled it with water, poured in a bottle of iodine, and left it on her desk.)

If any of us felt a higher emotion about our work, we did not freely express it. We still don't. I've tried to unravel that in my own mind because any idiot knows we had a special place, at a special time, in the life of this country. Could it reflect a subconscious fear that if we admitted this, even to ourselves, NASA might somehow discover they had the wrong guys?

In late July 1973, Al Shepard appeared as a guest on William F. Buckley's television show, "Firing Line." Naturally, they looked back over the years of space exploration and measured the worth of that effort. At the end, questions were invited from an audience that was mildly hostile—well-behaved but anti-space. One young woman asked Al if the astronauts had made a spiritual commitment to this "adventure" in which they were taking part.

I believe Al understood the question as having to do with religion—did the astros experience some kind of revelation?—and he dismissed it rather quickly. The question was valid and, watching on television, I reacted to it, almost eagerly. It went to the guts of what I consider another astronaut myth—the bland, plastic, programmed human machine myth.

There was, in fact, a spiritual commitment to adventure. To some of us it damn near became a religion. We were well aware that in the final third of the twentieth century, there were few frontiers left to raise the vision of man. From our unique vantage, space represented the last horizon, the last chance "to strive, to seek, to find and not to yield" until we had found another world. There will be other last horizons, but this one was ours.

Spiritual? Hell yes! I can't imagine anything more spiritual than the desire and drive and dedication to identify with the heavens, to achieve a oneness with the universe. All the emotion-charged phrases that people usually reserve for religion apply here. The job required that kind of commitment. You had to believe in it. It made you willing to risk your life, your health, your family, and even to lie about things that might kill you, just to experience the exhilaration of exploring whatever was out there. I think Al missed the point there, because if any of us reached it, Al did. But while we were in the program we didn't let such thoughts surface.

We weren't nearly as discreet about the kind of things that tended to chip away the hero's armor. We certainly didn't pay much attention to the words of Satchel Paige: "Go easy on the vices, such as carrying on in society. The social ramble ain't restful."

I had never owned and had rarely rented a tuxedo prior to arriving in Houston, but the old hands told us it was a high-priority item. That was an understatement. We were invited to so many social engagements, to so many charity affairs, to so many official (meaning, "by direction") functions, and into so many wealthy Houston oil-gothic homes that we would joke about it, insisting that the program needed a special category of astronaut to handle the night life. We would call them "partynauts."

The social life contributed to the impression that we were living the fat life, that the government was pampering us. But until the late sixties the extent of the government's largess for the astro execs was a per diem allowance of $16 a day while on the road. If it was worked carefully that could support an astro for a whole day in New York . . . if he slept in Central Park. The fact is that the astronauts did not get rich in the program, but we enjoyed a standard of living and social life that greatly exceeded our income.

There were constant invitations to spend summer vacations at private guest ranches, holidays at Acapulco resort hotels, and one private hunt after another. One aspect of this tempting schedule never stopped bothering me: as astronauts, we were usually the biggest dudes in the crowd—and frequently the only ones who couldn't really pay their own way. We were mixing with a crowd that was socially and financially way above us. Their standard of living had come to us through some kind of artificial insemination.

The new astronaut, of course, was like the little kid pressing his nose against the candy store window. He got an occasional lick of the peppermint stick but when it came to the bonbons, he usually went away hungry and envious. "Why don't I get invited to the posh affairs?" he might wonder. "Why can't I get some of the goodies?" We had received so much adulation, for the mere fact of being selected, that it was sometimes difficult to remember what we came to call John Young's Law: "One wasn't really an astronaut until he had been in space." Among those who knew, we were regarded as merely candidates. The guy they wanted for their parties, their balls, and their homes, was the astronaut who had been in space. Then they could ask, "Howazit? What did the earth look like from *up there?*"—all the things that made for good cocktail party talk. In Houston in the 1960's, nothing dressed up the decor quite like an honest-to-God astronaut. Potted palms had nothing on us.

But the law of supply and demand also applied here. When there weren't enough real live heroes to go around, the new guys would get their chance to fill the cocktail party conversation breach.

As astro pledges we would get such questions as: Is it hard to get in the program? Is the training difficult? Do you think you'll get a flight? What's Al Shepard like? To this day, thirteen years after joining the club, I encounter strangers and even old friends I haven't seen in years, who will ask with honest curiosity, "Do you *know* Neil Armstrong?"

It would have been much the same anywhere else, of course, but the fact is that most of it happened in Houston. What happened to us was flavored by the unique texture of that city, its oil riches, its friendliness, its love affair with the big deal and its admiration for those who make them.

And don't forget the Astrodome, Texas' answer to Xanadu. We had damn near carte blanche to whatever might be going on at the "eighth wonder of the world": "Have your secretary call, we'll leave your tickets at the box office." It was partly altruistic, an eagerness to do something nice for the local heroes, and partly just a public relations reflex. The pound of flesh would come when a message would flash on the Astrodome's $2-million scoreboard, "Welcome Astronaut So-and-so." Then there was the time when Lo and I and the Schirras were attending a concert at the Houston Music Theater—on passes. We were introduced in the audience and as we sat down Wally grinned and whispered, "We just paid for our tickets."

In a convoluted sort of way, NASA encouraged our socializing. We were selling the program, mixing with the community. Only rarely did anyone question whether the carefully contrived astronaut image was being well served as we made the rounds from Houston to Washington to Hollywood. Once someone approached Schirra at a party and asked if the drink he held in his hand might not unfavorably influence the youth of America. "My son is over twelve," replied Schirra, tartly, "and he's the Boy Scout now."

We were always being shown to the best tables and the best seats. Our first year in town Lo and I were invited to a charity performance by Marlene Dietrich. When we noticed the $100 price on the stub of our complimentary tickets we were stunned. In our travels we were frequently someone's guest at dinner—a politician, a government contractor, or a friend of the space program. That type of entertainment was accepted by us all and we never considered it influence peddling. The only risk involved was that some enjoyed it so much that they expected to be entertained and ceased being appreciative. That attitude help to turn off the faucet for those who came after.

Despite our modest $16 per diem, we were able to really stretch our money. We frequently stayed at the better places, where we enjoyed not a commercial or a military rate, but an "astronaut rate." We felt that we should pay something, so our hosts were willing to oblige our pride, and a buck or a buck and a half a night struck us as a nice democratic figure.

The more experience one gained and the wider his exposure, the more contacts he made. Favors and gifts, ranging from the casual to the substantial, became a habit. It began almost from the day we arrived. The big daddies of Houston opened their hearts, their homes, their offices, and their promotional schemes to us. When the Mercury astronauts landed, they were each offered a *free* home by a millionaire land developer destined later to become the center of the hottest bank and insurance scandal in Texas history. NASA, deciding that such an offer went beyond the bounds of friendliness, said No.

Soon the space agency was in the position of a blocking back, trying to fend the deals off. They kept coming, and though NASA mowed down quite a few, some got through.

Many offers, of course, were honestly made and seemed reasonably unselfish, and we accepted them. The insurance company that provided 4 percent mortgages on our homes never publicized it. It was one more way for me to live beyond my means.

Even before I left California to report to Houston, Al Shepard told me about the "brass hat" cars available from the local Chevrolet organization. These were management-driven cars sold to favored friends at approximately dealer cost, a price so sweet we could drive them for a year and sell them for nearly as much as we paid. We, the third generation of astros, were cut in because it was time to place orders for the new models, the total number was still small, and the deal was big enough to share the goodies. Besides, the parking lot would have looked funny with sixteen haves and fourteen have-nots. As it was, the parking lot looked like Chevy Town.*

Those were the first brand-new cars most of us had ever owned, and they had an immediate effect on our lifestyles. Corvettes blossomed like crabgrass, setting a pattern for years to come. I ordered a sedan even before I arrived, passing on the Corvette only because I already owned a Porsche. The new Chevy was waiting for us the day we touched down in Houston and I decided, "We're in Fat City. This I could get used to." We weren't cut in on everything, but enough to let us think we were insiders.

When I had learned we would share in the Time-Life and Field Enterprises publishing contract, to the tune of an extra $16,250 for the next several years, Lo and I began to reshape our plans for a modest home in a community adjacent to the space center. Within weeks we went through the exercise of moving from a home in California that had cost $14,500, to talk of buying a new one for $25,000, to consideration of building one for $50,000. Admittedly, that was a wild dream for a rookie astronaut making $13,000 a year, but I went for the better home, gambling on an ever-improving future. It wasn't a sure thing, but it was a helluva good bet.

If we rationalized the favors we accepted, it was on the basis of the most fundamental of all reasons: money. We felt a desperate need to break even and still meet our obligations in the cage formed by our

* We went bipartisan, Ford and Chevrolet, a few years later when our ranks doubled.

media image and NASA obligations. We were expected to travel all over the map, look great, talk great, act great, and all on $13,000 (or less) a year! To varying degrees it was a conflict that unsettled every man who came into the program. Gordon Cooper once explained it to a writer from *Esquire*: "Being a celebrity is something new to all of us, at least on this scale . . . it isn't always pleasant. We were being used as a kind of shot in the arm to the nation." And Gordo added, "The *Life* magazine money is all that has kept our heads above water."

Poor *Life*. I doubt that anything they ever printed about us got more publicity than the contract itself. And poor Field Enterprises. By the time my group got in on the action, Field was shelling out nearly twice as much money as *Life*, but everyone still referred to it as the Time-Life contract. By whatever name, it was, as Cooper said, a godsend to the astronauts. Without it the temptation to moonlight would have been greater. Rather than contributing to the grubby scrambling that came later, I think it actually held it off till the contracts had expired.

From our point of view, of course, it was a nearly perfect deal: no reporters hounding us, $100,000 in life insurance provided by the two publishers, and a cash bonanza that in nearly every case exceeded our government salaries. And, best of all, there was little we had to do for it.

The contract signed in 1962 was for four years with an option for four more. At first each astro was assured $10,000 a year from Field Enterprises plus $6,250 from Time-Life. However, the contract called for a maximum total payment per year, distributed equally among all the active astronauts, and the last year anyone received the maximum amount was 1965. Overnight it looked as though each of us was worth an extra $100,000 over the next eight years—*if* the option was picked up *and* the group didn't grow. As a matter of fact, the participants grew to seventy-three and the option was not exercised by Field Enterprises.

But a significant part of all this, one that was consistently overlooked in the debate over the ethics of our accepting money for stories some felt already belonged to the public, was the insurance coverage. The cash was fine, and most of us liked it only a little bit better than our right eye, but at that stage in space travel it was impossible to find an insurance company that would write a policy on our lives. Ironically, it wasn't long after we joined the program—that is, the third group, the Apollo astronauts—that the publishers had to pay off $100,000 death benefits on the lives of Ted Freeman, Charlie Bassett, and Elliott See.

Field Enterprises was particularly generous in their administration of the contract. Even after they had elected not to pick up the four-year option, they continued to pay the widows the full contract benefits as though it were still in effect. It was a no-strings gratuity on the

part of Bailey Howard, the chairman of the board of Field, and reflected the generosity of the entire Field organization toward the
astronauts.

In some ways, the contract haunted everyone who touched it. I don't
think any of us ever felt that the stories—those by *Life* or Field Enterprises World Book Science Service—did justice to the situations they
were intended to describe. We always came out bland, very plain
vanilla. Of course, every story had to be approved by NASA, and
that arrangement discouraged anything that looked interesting or
spicy.

The writers, all of them, made it as convenient as possible. In a sense
they worked for the house. They were friendly and openly rooting for
us. Yet time and again they were writing the same stories, under the
same conditions, and it was inevitable that we would turn out to be
almost indistinguishable from one another. As Robert Sherrod, one of
the *Life* writers, was to put it later, "The astronauts came out, as usual,
deodorized, plasticized, and homogenized without anybody quite in-
tending it that way."

We weren't too satisfied with the results but we salved our critical
perspective and our consciences with the money. The publishers were
less than thrilled with the results, and pleaded constantly for more
access to the astros, better insights, more honesty, more exclusivity.
Various books have been published from the material, including *We
Seven*, about the original seven astronauts, and *First on the Moon*, the
story of Armstrong, Aldrin, and Collins. Both bombed.

The Mercury astronauts received approximately $25,000 a year for
the duration of the Mercury program from *Life*, which turned out to
be the only serious bidder for their personal stories. By 1962, when the
second, or Gemini, group of astros had been selected, President Kennedy had decided that the contract would not be renewed. The ethics
of the whole greasy deal bothered a lot of people but most notably the
press, whose sense of moral outrage is always heightened in situations
involving its own interests.

The authority for the astronauts to sell their personal stories dated
back to the Eisenhower administration. Before the controversy ran its
course, just about everyone had taken sides; the press, the public, right
on up to the President of the United States.

John Glenn, who had become a particular favorite of the Kennedy
family, probably saved the contract. One of the arguments advanced
against the sale of our private stories was the obvious one: soldiers in
combat also take grave risks, but don't, as a rule, get paid for talking
about it. When the president asked John about his feeling on the
decision to shut off the *Life* contract, he cut quickly to the point. "I did

not deny the old argument that a soldier going into combat might share an equal danger with astronauts," John later explained, "but I felt that if there was enough interest in that soldier's home life, background, or childhood, then he, too, should have the right to receive compensation for opening his home, his family, and his innermost thoughts to public scrutiny that would not otherwise be available."

Although by 1970 the payments had dwindled to less than $3,000 a man, and in 1972 it was $64, the contract was one of the few things in which the astronauts shared equally. Other windfalls were a different story. Some astros had what the marines facetiously call the Semper Fi attitude: "Pull up the ladder, I've got mine."

No one was handing out uncut diamonds or GM stock but Gordon Cooper, for one, was always willing to share, whether it was free Firestone tires or arranging for everyone to own a diving watch. Tires, watches, cameras, and other assorted items came our way occasionally. Some passed them up but a few freeloaded on everything that came their way.

Wally appropriately labeled the activity "scamming" and those astros who were really good at it came to be known as "scam artists." Wally for a long time had the dubious distinction of being one of the best.

I doubt that any of us ever agonized over any deeper implication of the favors, gifts, and services that were lavished on us over the years. It simply meant that America loved us, and our wives. It would be ridiculous to presume it could influence anyone's judgment; besides, we were never involved in contract awards.

For the lift-off of Apollo 7, I wanted my family to see and understand what it was for which I was willing to invest five years of my life. And I wanted them to feel it emotionally, not just read about it. Since I hadn't broken myself completely of the old-fashioned habit of paying my own way, when NASA wouldn't provide my family government transportation—they couldn't (or wouldn't) draw a line between an astronaut's wife and those of other NASA officials—I automatically booked them on a commercial flight out of Houston.

When I asked a friend to meet them on arrival, he was appalled—I mean, shocked—that the flight crew's family had to fly to the Cape like everyone else. Over my protest, mild though it was, he picked up a phone and arranged for them to be picked up by one of the many corporate planes headed for Florida. Every Apollo flight drew a veritable air force of private jets owned by corporations and business honchos and Texas oilmen. Many of these aircraft were only partially filled and the owners were glad to be personally involved with the mission and its participants. Lo (and sometimes the kids) have traveled to or from the Cape on airplanes belonging to Rockwell, Ryan

Teledyne, Howard Hughes (courtesy of Bob Maheu), and many others.

Although the wives had a lot of legitimate complaints about the neglect and loneliness that resulted from their husbands' being totally wrapped up in their jobs, they shared in many of the perks as well. They grew accustomed to the privileges and fringe benefits of the program. They got to fly around the country in plush, private jets. They were the center of attention at club meetings and social functions they attended. In general, they were treated like queens. Some wives found that life harder to give up than the astronauts themselves.

The endless Las Vegas jackpot spitting out all of these good things only created more problems at home. It was a monster that had to be fed. As our standard of living rose, the pressure for a greater cash flow grew. It generally fell to our wives to cope with the monthly budget. The independence they developed would sometimes further agitate the troubled home life of a family separated half the time.

As each astronaut became more embroiled in the supermarket of life—it was like one of those shoppers' sweepstakes on TV where the wife races up and down the aisles pushing a grocery cart and grabbing meat and canned goods against the clock—the more tempting seemed the business offers that others kept dangling in front of us.

No, not every business acquaintance was out to exploit us, but it paid to suspect that he was. And there was more than one way to exploit an astronaut.

On the one hand, he could know that he was being used and not mind it, because the situation provided what he wanted in return: income or contacts or the promise of a future connection in the business world.

The easiest way to fall into a cesspool of trouble was to be naïve enough to always think that they—the big operators—really needed us. That they admired our brains, not just our bodies. That astronaut hero image was directly convertible into dollars if handled right, but no one wanted to be obviously exploited. We pictured ourselves as astute and preferred to believe we were solicited for that reason. Actually, it was nearly always because of our commercial value. The deals usually came to us sugar-coated in order to sell us on their program.

Several years back Alan Bean became a director of an insurance company in what happened to be a really fine deal. Bean is cautious and he negotiated for a nice piece of stock at a very friendly price, on borrowed money, at brother-in-law interest rates. A few months later he improved on the original deal and it looked like he was on his way to making a killing. When I last heard the stock was down, he still owned a lot of it, and it appeared if he cashed out he would be a loser.

In the meantime, the company had made use of Alan Bean as a director and more. He had conducted the grand tour to company principals at Houston and the Cape, and entertained them at his home with the astronauts. They have been able to parade the "company astronaut" when necessary or useful. Another astro business deal that didn't strike gold.

The older hands were always cautioning the young astros against being too gullible, with a dash of oneupsmanship or gamesmanship liberally applied here and there. Guys like Schirra or Stafford might badmouth a deal simply to give the impression that what you had working wasn't nearly as good as theirs. And generally it wasn't.

None of this, I hasten to add, was illegal or even shady. We weren't running guns to the Middle East or filling orders for the fleshpots of Cocoa Beach. But unless one thought the whole world was a cream pie, he had to know that when a company tacked your name onto their board of directors, they intended to use you. Few of the guys had any real business training, and almost no one had experience in the commercial world. But we knew our reputation had value and figured we should be getting a quid pro quo.

Looking back, it is remarkable how many of those early associations ended in disaster. Almost all of us, at one time or another, wound up getting singed. And I can't say we weren't warned. There were regular Civil Service rules limiting the business activities we could engage in. In addition, the NASA administrator distributed a three-page letter early in the game listing further restrictions on astronauts.

We received a fairly steady stream of the sky-is-falling memos that tried to alert us to the dangers. In essence, we were not to accept part-time jobs, consulting or lecture fees, or gifts, and we were not to involve ourselves in any way with companies competing for government contracts. NASA was concerned not only about what we were doing but what we might *appear* to be doing. Human nature being what it is, the warnings and restrictions did not eliminate all such activity. That's one of the reasons adultery is so popular, because it's illegal.

Some of the boys were amused by the fact that the virtue-and-purity lecture became the duty of Al Shepard, then chief of the Astronaut Office and in the mid-sixties on his way to becoming a millionaire in banking and real estate. Actually, there was nothing in the NASA rulebook that said an astronaut couldn't invest in business, and even participate, so long as it didn't involve huckstering. When his ear problem grounded him, Al quietly used the contacts he had developed in Houston to build a handsome investment portfolio. Through friends in auto dealerships, banking, and the construction business, Al became a partner in a couple of banks and an investor in shopping centers and discount houses.

Al is a little sensitive to the suggestion that he got rich on NASA's time. But he was able to handle his office responsibilities with about half his energies. He never stopped fighting to get back in line for a flight, and when it came—Apollo 14—he shed most of his holdings and all of his directorships. Of all the Mercury Seven, Shepard's career was the only one aimed well enough to get to the moon. He stuck it out. He survived. I don't doubt that Al would have made a pile of money in or out of the program.

At any rate, neither the memos nor Al's lectures really discouraged us. We felt that NASA was intruding too much into our private lives, including our precious American right to be screwed by some hard-nosed business promoter. It was the old story of Big Brother trying to protect Junior, only once the kid has been to town and seen the bright lights he's not interested in being protected.

Part of the romance of a big deal is being involved with the high rollers. We found ourselves moving with the likes of John King and Clement Stone, Robert Maheu and other men who controlled sprawling business empires at the time. But the high rollers weren't immune from catastrophe either. King lost control of his own King Resources and Maheu was fired by his invisible boss, Howard Hughes, in a weird sequence of events that created slightly more publicity than the first moon landing.

In the early days it hardly occurred to us that millionaires were also subject to wrong guesses. We learned quickly that if we gave it the personal touch, turned on the astronaut charm, and touched their lives, it was easy for such men to become enamored of the space program. They became fans and space buffs. Casual contacts with influential people could deepen into friendship but it generally started with that first impression, that big flash, of getting high on the astronauts.

I submit with some regret my first extracurricular business involvement as a classic example of the whole process, better than some in our office and worse than others. It began with the purchase of stock in an electronics company on the recommendation of a lady broker who, it later developed, was dating the president of the company. One of their products was a briefcase telephone, an item that never quite caught on with the masses. The stock was rising, and at the urging of the broker I met with the president of the company, one of the original founders of Litton Industries. He was impressed, he said, by my background, my technical knowledge, my charisma, and, of course, my good judgment in having bought his stock. He invited me to become a director and a consultant, at a retainer of $1,000 a month. That was roughly the going rate for astronauts in the corporate marketplace and wasn't an unusual consulting fee.

This was my first brush with the commercial corporate world, and as near as I could see (not far past the $1,000 per month) the deal looked good. I began unraveling the NASA red tape required to affiliate myself with the company, making the usual declarations that they wanted me because I was such a smart apple and not because I was an astronaut. I also warranted that there would be no exploitation of my position in the company's advertising. The application had to be approved at three levels, up to and including the NASA Administrator, James Webb. (Such approval was required even if no compensation was involved. To become an unpaid director of the nonprofit Earth Awareness Foundation, I had to go through the same process.)

Though it wasn't made easy for us, NASA tried to not shut the door completely and, except in those cases where a conflict of interest clearly existed, the applications usually went through. But the agency could have said a flat no to any outside activity. And, given the big club it carried—control of the flight assignments—most of us would have gone along with a minimum of kicking and screaming.

I got my approval, happily pledged my allegiance to my new business "friends," and plunged right in. The company was based in California, and since I was flying back and forth regularly to North American Rockwell, it was convenient for me to stop in occasionally to shake hands or catch a directors' meeting or offer a little advice.

The relationship proceeded so smoothly that I even received a check for the first month's consulting services. Then I agreed, in lieu of salary, to accept stock which would eventually turn out to be worthless.

I had originally purchased several hundred shares of the company stock on the open market. As luck and destiny and my own sense of timing would have it, my one good shot at getting out with a fat profit came during the flight of Apollo 7. The stock doubled in price reaching its all-time high during our eleven days in space. Several of the astros had bought the stock, including my crewmate, Donn Eisele, so we worked out a code through Jack Swigert, one of our CapComs, to pass on the latest quotations. As the price kept rising, Donn and I squirmed, trying to think of some creative way to tell Jack to pass the word to our broker to sell—we never did.

It took six months to realize most of the directors were no better equipped to run the business than I, and less prepared to learn. The company began to flounder. The stock fell, and I watched in vain for signs of improvement. As financial problems mounted, they came up with a private placement, with the stock offered at "bargain prices." A number of the fellows jumped in with me on the offering—Armstrong,

Collins, Bean, Kerwin, Weitz, Anders, Schweickart—even though I pointedly made no effort to tout them on the stock. But with fine logic, they figured ol' Walt was a director of the company and assumed I wouldn't be there if it wasn't a good deal, so they bought in. It left me feeling like I had sold a used car to a friend, and I kept my fingers crossed that it would all work out.

Next the company became the target of a take-over attempt. An elderly woman, who held a major block of shares, was being romanced by an equally elderly flim-flam man, then under indictment in Los Angeles in connection with another company he owned.

To attack the present management he tried to get at me, planting tips with the television networks and several newsmen that I was a "front man for a bunch of crooks." It was an uncomfortable time. Even though none of the stories was ever aired or printed, I felt trapped.

It took eighteen months for me to wend my way through the following scenario. Phase 1: "Well, I'm just a neophyte at this, I'll sit here in the board room and see how the big boys in business do it."

Phase 2: "My God, they don't know any more about what ought to be done than I do."

And Phase 3: "Are they honest? I really shouldn't be involved with these guys, but since my friends and relatives have invested I'll hang around a little longer to see what I can do."

Finally I resigned, and a few months later the company was in bankruptcy.

I wound up with ten thousand dollars invested in worthless stock, plus a bucketful accepted in lieu of salary, plus the pinched conscience of knowing that friends and relatives had blown some dough as well.

I was not the only astronaut to take such a bath. For men so astute at their jobs, with reasonably good heads, to be taken for such rides in business, can only be attributed to the most naïve assumption of all: we trusted the public not to screw an astronaut. In the late sixties this was the cornerstone of many of our business deals.

There was Shepard, operating out there with the captains and kings of industry, and here were the rest of us, losing our lunch money. Tom Stafford was teased about his bad oil deals although he actually only invested in one dry hole. Jim Lovell had his name dragged into Texas' largest bank stock fraud. Frank Sharp, the principal in that one, even promised to make Jim the President of Braniff Air Lines, except for one small detail—he forgot to buy the airline.

Shepard, Slayton, and on up the line must have shuddered each time one of those special requests hit their desk. The topper came when Buzz Aldrin, catching up on things after Apollo 11, sauntered in one day and plopped down on Chief Astronaut Al Shepard's desk a whole handful of requests to become involved in outside business deals. After that, the agency began to take a harder line on the nature and number of such activities.

Another favorite sport was the stock market.

Occasionally some insider would attempt to ingratiate himself by giving one of us a hot tip on the market. Incredibly, the first five years I invested in the market I never lost a cent. Either I was always listening at the right time or the market was fixed. Whatever, the next four years I never picked a winner.

None of us were what you might call big plungers. If I heard about a stock selling at 5, and I bought a hundred shares and sold at 8, I had made a score, $300, and I was damn glad to get it.

Typical of the way we operated was the tip that Roger Chaffee passed to several of us in 1965 on a company called Liquidonics. Couldn't miss. Roger already had his little edge. He bought it at 4, tipped us at 5, and sold out at $7\frac{1}{2}$. I dumped mine at $7\frac{3}{4}$, on the assumption that Roger had to know what he was doing. And besides, I had made my $200 or so. Liquidonics continued all the way up to 70 and, though none of us exactly ate our hearts out, we did chew a little on the left ventricle.

You could walk down the hall of the astronaut offices and find behind each door a desk piled high with work and an astro—if he was in town. We all seemed to have a phone growing out of our ear and if it wasn't space business, there was a good likelihood we were talking to a broker or lawyer or to some home-state politico about a personal appearance or opportunities on the home-grown political scene.

As an example of this hectic activity, in early January 1969, immediately following his flight around the moon, Bill Anders and I were walking toward his office. As we passed his secretary's desk he called out without even thinking, "Jamie, get Dirksen on the phone." It seems that Everett Dirksen, the Senate Minority Leader, had been trying to reach Bill for several days.

Bill's casual remark, as he stood there in pinstriped suit with vest, struck me as totally incongruous with the Bill Anders I had met five years before. He had come a long way from the "cruddy shoes" Anders selected as an astronaut in 1963.

Looking back on the heyday of Apollo, I marvel at our collective innocence. An astro would return from a space flight. He would be invited to his home state to speak at several banquets. He would take the standing ovations, hear the cheers, and gaze out upon those adoring faces. Local politicians would see the tremendous crowd appeal and suggest that he run for office. In his head he would suddenly be hearing "Hail to the Chief" or something. It was the siren call of "public service," and the victim would think, Why not?

The irony of it was that most astronauts were military men and had lived for years under the Hatch Act restrictions, which prohibited political activity. They avoided national politics, considered themselves transients, seldom took part in local issues, and were usually absentee voters from their own home districts. Most of them were politically unaware and could be described as apolitical. But they had been exposed to the great white light in Washington. They had paraded before Congress to defend the budget and sell the space program. The smell of the power and the glory was a strong attraction.

This astronaut-as-political-prospect began with John Glenn.

By the time I reported to NASA, it was already rumored that Glenn would return to Ohio to run for the Senate seat held by Stephen M. Young, a seventy-four-year-old Democrat. John was seen as a handpicked Kennedy candidate. He was the talk of the Astronaut Office, and the bitterness I sensed among his old Mercury teammates was disturbing to me. I learned later that it stemmed from two causes. John had been the program's straight arrow, pure of heart, the fellow who kept insisting that astronauts had an obligation to the American public not to misuse the fame and power that would surely come their way. We had to set examples. Well, some of the boys wondered if the Senate was where you went to set an example.

I'm sure they felt that John had let them down, that he had capitalized on his stature as an astronaut to launch his political career.

There was another aspect of John's ambitions which bothered them. After that first orbital flight, the country—and the White House— simply adopted John Glenn. Of all the astronauts, he was the one with the most magic goofus dust on his shoulders. While the others were off training for the next Mercury missions, and probably bitching a little about the work load, they could pick up the paper and see photos of John and Annie Glenn skiing with the Kennedys, weekending at Hyannisport, or riding the rapids somewhere with Bobby. Scott Carpenter and Wally Schirra felt he should have been at the Cape helping. How much of their feeling was fraternal and how much was simply envy, I couldn't tell. Before it became obvious that the political bug had smitten John, they honestly worried about "the team" breaking up. Wally Schirra was the one who brought it into the open, declaring in a television interview that John's public relations activities had just about washed him out of the space program. He was hopelessly behind.

At the beginning John was about as political as a Saint Bernard. Though he had been given the Kennedy liberal stamp, he was like most military guys, at heart, conservative. I even caught him once with a copy of *Human Events*, a conservative organ, in his mailbox. John simply believed in the old-fashioned virtues: flag, honor, a strong military, and short hair. You can know a career officer all his life without ever being aware of how he votes or if he does. It's like when Eisenhower came home from Europe to run for president. His platform was his grin. No one knew for sure what he was; and he could just as easily have run as a Democrat than as a Republican.

With most military men there is room for growth, and this was true of John Glenn. By the time he made his second race for the Senate in 1968, he actually believed in and supported the liberal issues that his backers had assumed he favored all along. He had developed his own political consciousness, and it had been shaped by those around him: Bobby and Ethel, and McBundy and Sorensen and the rest of the Kennedy crowd.

None of his Mercury fellows could get involved in the campaign and they were probably relieved by the restriction. Rene Carpenter did go to Ohio to campaign for him. I suspect the others were gun-shy and nervous about what such political exposure would mean to them in terms of the program; they were inhibited by all those years of military political isolation. They also must have resented John for making them feel politically uncomfortable.

It really wasn't so surprising that he made the decision to enter politics. When he flew the first U.S. orbital flight he was pushing forty, and he knew the odds were against his making another. As for landing on the moon, that was way out of range. Given his clear status as the crowd pleaser, the program's Number 1 hero, there was a feeling in the establishment that John shouldn't take any more risks. As he himself finally put it, "The idea of becoming the world's oldest, used, permanent-training astronaut does not strike me as very good career planning."

John was the tip of the iceberg; then the floodgates opened. Both parties became aware of us as a virgin talent pool, would-be candidates for national office who came with something almost as valuable as a name or money or a famous face. We had charisma.

You could sense it as a guy finished postflight activities or moved into his late thirties. All over the place astros were preening themselves, posing on imaginary runways like beauty contestants, waiting for the phone to ring.

To take the fruitcake, those who saw themselves as attractive (irresistible) candidates, all wanted to run for the Senate! They weren't your future sheriffs and city councilmen and justices of the peace. No, sir. These were unselfish men, and if the country needed them to serve in the United States Senate they would not disappoint the public.

There was always scuttlebutt around the office or in the gossip columns about this one or that one considering a plunge into politics or being courted to run. Al Shepard, who may honestly be the only one of us with no such ambition, was sought after in both Texas and Virginia.

At one time Tom Stafford was all but barnstorming across Oklahoma, trying to make up his mind about a race for the Senate. Jim Lovell was the subject of a constant stream of rumors out of Wisconsin. Jack Swigert "confided" that he had been approached to run in Colorado (for the Senate, of course) as was Charlie Duke in North Carolina. Don Lind, a Mormon, was touted as a prospect in Utah.

Who is Don Lind? A good question that illustrates the depth of the astronaut charisma. Lind is a civilian scientist-astronaut, with ten years of experience and scant prospects of ever lifting off the pad. He came into the program in 1966 as part of the fifth group selected—which had dubbed itself, tongue-in-cheek, "The Original 19." He was pleasant, bright, a naval reserve aviator with better scientific credentials than operational flying experience. He is a classic example of how promotable the image was. If the public were to rank astronauts on accomplishment and recognition—like football's top twenty—Don Lind would be listed among "the others." Not a glamor boy, he had labored more than ten years without making a flight. And yet, in 1971, key people in Utah were encouraging him to run for Congress. As far as his co-workers could see, he had two things going for him: he was an astronaut and a devout Mormon.

In a very real sense, Lind was "Utah's astronaut." That was one of the quirks of the program. NASA had a penchant for sending a guy back to his own state for speeches and diplomatic errands, and naturally we did nothing to discourage it. In some places it magnified territorial pride into something really fierce. (For an occasion I've long since forgotten, I once made a banquet speech in Milwaukee, Wisconsin, Jim Lovell's home state. When I was introduced, I simply said, "I bet you didn't know NASA had more than *one* astronaut," and the place collapsed.)

All of which led to a little disappointment around the office. Some bright opportunist would listen to all the political gossip and measure his own qualifications and conclude he was both electable and draftable. Then he would wonder why a group in Wisconsin was trying to run Jim Lovell for office when his own state had yet to realize they had this hot property waiting by the phone down at Clear Lake, Texas.

For all the maneuvering, real and imagined, that was going on, we engaged ourselves in precious little political dialogue. This was due to military conditioning as well as discretion in getting involved in matters NASA might regard as too controversial. No prohibition of it, just the ever present sword of crew assignment hanging over our heads.

In 1964 I was one of two astronauts openly supporting Barry Goldwater in the presidential race. The other was Elliott See, who wore a "Pilots for Goldwater" pin on his lapel and caught some hell for it. Goldwater had not been a supporter of the space program and that worried most of the people at NASA, even those who identified with his other positions. Favoring Goldwater wasn't a popular thing to do just then, but between being a civilian, a reservist, a pseudo-scientist, and a fresh rookie, popularity wasn't going to be my strong point.

The partisan conversation was sterile and the political activity almost nil with a few exceptions. On the morning after Martin Luther King was assassinated, Buzz Aldrin was leading a memorial parade in downtown Houston. Under the circumstances it was a nervy thing to do, and Buzz caught a lot of flak for it and no end of criticism from the rest of us—behind his back. It was mistakenly assumed by the public that he represented the astronauts when he only represented Buzz Aldrin.

But by and large no one knew what the astronauts stood for (other than God and country and the space program), sometimes an asset in itself in politics. There was also the clean image and the fame and romance that went with being space heroes. It's ironic that no one effectively harnessed all that. Fifteen years after the first Mercury flight, only two have managed to get elected: John Glenn on his third campaign for the Senate and Senator Harrison Schmitt on his first try. My own flirtations, I must admit with a slight blush, were set only slightly lower. My political ambitions grew out of my career as an astronaut, and my goal was Congress. I was a conservative civilian, politically informed, and willing to take a stand and argue a position.

The kind of show business glamor that attached itself to the astronauts had never been a handicap in my home state.

Years ago I planned my life around a loose timetable: my twenties were devoted to education and my thirties to space flight; in my forties, I would work to provide financial security for my family, and in my fifties, if I could afford the luxury, I would consider going back into public service. Politics almost rearranged that schedule in 1970.

When word began to circulate that I was thinking about leaving NASA, I was approached by several influential people about running for Congress from California. The attention was flattering, and the notion of running for the House of Representatives seemed more appropriate for a beginning politician than a bid for the Senate. I was immediately faced with the same question as a Hollywood actor who is criticized for taking a shot at politics; there's glamor and a name but what about the know-how and everything else it takes?

It's a legitimate question, and the only answer I've heard that made sense came from John Glenn: "Why not, I'm honest, I'm intelligent, I'm capable."

I had the same attitude. The country needs honest men, who can resist pressure and who are reluctant to compromise what they believe in.

An astronaut establishes some valuable contacts. From mine the encouragement was widespread, firm, and consistent. Redistricting had created five new seats in California, and I was intrigued with the prospects in my old home district around Santa Monica, where the incumbent seemed to be in trouble.

I had delivered newspapers on those very sidewalks, worked at the local RAND Corporation, was considered something of a hometown hero, and would have the backing of the local newspaper.

I backed off simply because I did not want to run as an astronaut. I could picture myself forced to exploit the space image to a far greater degree than I was prepared to do. It would require compromises in order to raise the finances and money has never been a good enough reason for me to modify my beliefs.

I haven't abandoned those political ambitions but waiting till I can afford it still seems the best idea. I'll reconsider public service when being an astronaut is maybe the brightest of my accomplishments, but not the only one.

The fact that there has been so much sound and motion about astro political careers, and so little substance, suggests several things. It might have been our own good judgment winning out. By training, astronauts weigh their chances and don't take unnecessary risks; they calculate the odds against their own capabilities and train it down to a tolerable risk.

Again, those who approached us may have been more astute—or more cautious—than we gave them credit for being. It takes a lot of right answers before those with the pull and the money throw in their support. Perhaps they discovered that "their" astronaut didn't have really strong convictions. Or maybe they were looking for a man who could be steered, molded, and controlled, and that isn't an astronaut's strong suit, either.

Don't be surprised if several more of us end up in Congress, and maybe a few statehouses as well. The astronaut charisma won't disappear altogether. There isn't a one of us who hasn't learned to stand up in front of strangers and make an effective speech, the kind that brings them to their feet or brings a little puddle to one eye.

Who knows whether the hero worship, the romancing, the gifts, and free plane rides led to later embarrassments. I can't be sure. Maybe collectively we crawled along, taking what came our way, until a really motivated guy like Dave Scott reached a point where selling envelopes he had carried to the moon didn't seem wrong.

It is easier to get a perspective on the public's reaction. NASA went so far out of its way to paint an image of us as clean and pure, with hearts made of precious metal, that however we slipped it would seem like a fall from grace. As the saying goes, watch that first step. It's a lulu.

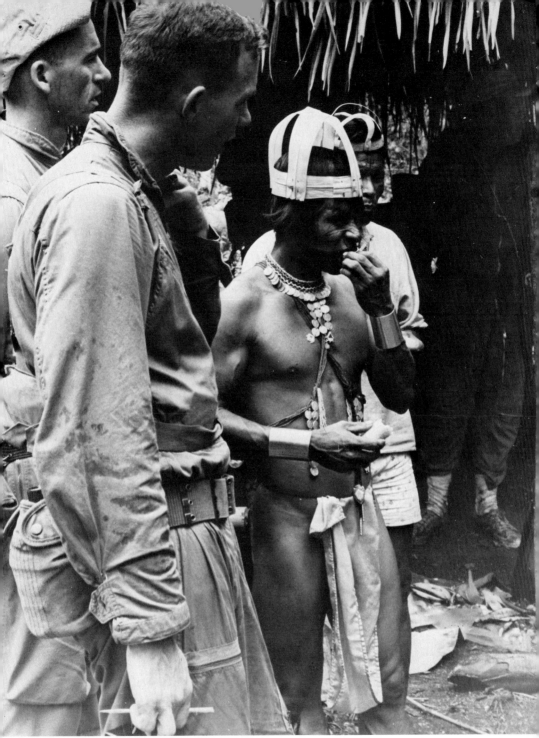

At the Tropical Survival School in the Panama Canal Zone, astronauts Scott and Cunningham watch a Choco Indian eating a wild banana at his jungle home. (NASA)

The Sheik of Araby? No, it's Walter Cunningham at the completion of training at the Desert Survival School in 1964. (NASA)

During a break in Apollo 7's flight crew training in the mission simulator. Left to right: Schirra, Eisele, Cunningham. (NASA)

The Apollo 7 crew at practice in the Gulf of Mexico. Coming through the hatch is yours truly. (NASA)

Entering the Neutral Buoyancy Tank for Skylab zero-gravity evaluations. (NASA)

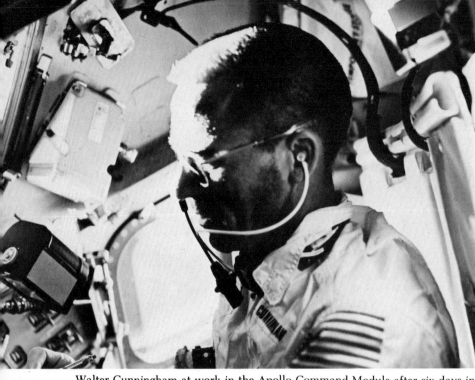

Walter Cunningham at work in the Apollo Command Module after six days in orbit. (NASA)

My wife Lo is holding Kimberly as they watch for Apollo 7 to pass over Houston in the early morning darkness. (Othell Owensby, *Houston Chronicle*)

Bearded Apollo 7 astronauts (left to right: Schirra, Eisele, Cunningham) emerging from the recovery helicopter which carried them from their splash-down point to the deck of the USS *Essex*. (NASA)

The Apollo 7 crewmen were guests at the LBJ ranch for the Apollo 7 post-flight press conference. (NASA)

We presented President Johnson with an Apollo 7 on-board photograph during the press conference and awards program at the LBJ ranch. (NASA)

The Houston parade for the Apollo 7 astronauts was typical of the many attended by the entire Cunningham family.

Two Soviet cosmonauts enjoy a social gathering with a group of astronauts during a visit to Houston. Standing (left to right): Armstrong, Aldrin, Maj. Gen. Nikolayev, McDivitt, Cunningham, Stafford, Swigert, Schweickart, Scott, Lovell, Slayton. Kneeling (left to right): Anders, Conrad, Vitali I. Sevastyanov, Gordon. (NASA)

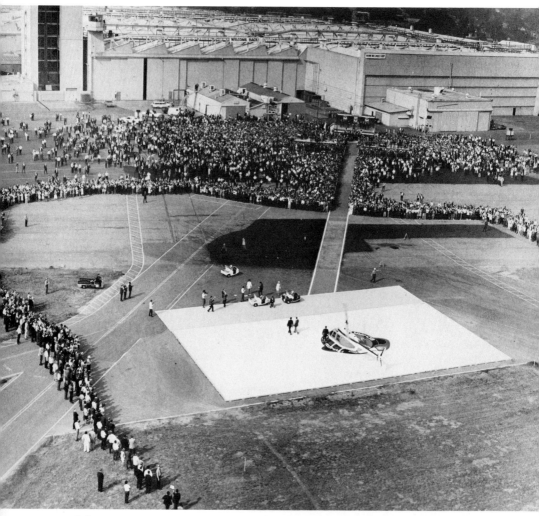

Schirra and Cunningham have just stepped off the helicopter in their first return visit to the Rockwell International plant at Downey, California, after the flight of Apollo 7. The Apollo Command Module was designed and built by the waiting crowd.

11

Don Juan the Astronaut

IT WAS APPARENT that the astronauts had truly arrived as national figures when, in the spring of 1974, a letter about our sex lives appeared in one of America's most widely read columns:

Dear Ann Landers:

I am curious about a certain aspect of our space program that has never been discussed publicly. Will you try to lift the veil?

Since our society has become so sex conscious, and any married man who abstains for an extended period of time is considered an oddity, what measures are taken, if any, to minimize or curb the astronauts' desires?

Their "sexual sabbatical" should serve as an incentive to wayward husbands who find it necessary to be away from their wives for extended periods of time. It might help them reassess their motives before they decide to stray.

Not in Orbit

Ann Landers replied, almost indignantly, that not all men who travel are faithless, and she went on to develop this interesting theory for two paragraphs. But the question went begging. "As for the astronauts," Ann concluded, "only THEY can answer your question. Anyone in Houston want to respond to the lady's question?"

As far as anyone knows, Not in Orbit is still waiting. For what it's worth, the answer is a simple one: Nothing was done to curb our desires. However, we probably thought less about sex then than at any other time. There was little opportunity and, as the scientists might put it, no stimuli.

But that letter wasn't to be laughed off. It brought to the public's attention the conflict between our Boy Scout image and the party-time gossip. The first clue we had had that our night lives might be grist for the rumor mill was when we listened to Deke Slayton deliver his "Be a Good Boy" lecture. His main result was to convince us even more that the job offered fringe benefits yet unrealized.

In early 1964, as the fourteen Apollo astronauts prepared for their first visit to the Cape, Deke called us together and cautioned us to go easy on the social vices. His immediate concern was the recent arrival in Cocoa Beach of a correspondent for a keyhole rag called *Confidential*.

"You guys are all big boys now," Deke began, as I recall, "and I'm not going to tell you how to behave, but I will point out a few special pitfalls. Most of you are bright enough to know that if you are going to be screwing around you'd better be damn discreet about it. Most of you, I hope, won't need this lecture. But all of you are going to get a lot more play from the girls now, and you're going to find that some of the habits that kept you out of trouble before won't save you here.

"Apparently, the people from *Confidential* have heard that we swing pretty loose and easy around the Beach. We hear they've now got a man watching the lobby of the Holiday Inn, seeing what they can dig up for a juicy story. They're also watching the back roads at night to see who is coming and going. If you think they'll ignore you because you're new, forget it. So if any of you have hanky-panky in mind, you'd better also think about this: we didn't bring you into this program to be smeared all over some scandal sheet. Any time your outside activities are more important to you than this job, just let me know and we'll arrange for a change."

No one spoke. Deke was to make it clear, time and again, and not just on the subject of chasing girls, that we were expendable. It wasn't so much a threat as a fact of life. We were all selling our services and NASA was the buyer. They didn't just own us from nine to five, either. They bought us hat, ass, and overcoat, and it was a buyer's market. The fear that they could and would cut us loose before we had flown a mission was what kept us in line. The ever-present threat didn't prevent us from engaging in certain irregular activities, but it did take a little of the fun out of it, and it sure kept us on our toes.

What wasn't realized at the time was the particular pedestal on which the media had placed us. In the years ahead they would contribute to protecting our reputations in a manner usually reserved for national political figures.

Over the years astronauts acquired a kind of underground reputation for bedroom athletics. It never really broke into the open; couldn't damage the White Knight legend, you know. Stories circulated about wild parties (orgies), and the names of some were linked with well-known actresses and entertainers. A brief epidemic of divorces about the time the moon flights were ending raised more than a few eyebrows.

There were some who led the kind of life Hugh Hefner might envy. And there were others who stuck by their marital vows and subsisted on milk and apple pie. But let's face it, we had a lot more opportunity to go astray than the average Rotarian.

What those around us should have understood from the beginning was that, essentially, we were fighter pilots. The observation of one friendly psychologist early in the space program was quite valid: "They really burned the candle at both ends. They had a tremendous vitality. They would work hard, train all day, and party all night. They did a lot of partying. They drank. I'd try to follow along on some of their activities. I couldn't keep up. I'd just pass out." Aviators have a saying: You can always spot a fighter pilot. He's the little guy trying to cash a check at the bar in the officers' club, with a huge watch on his wrist, ready to swing with the nearest good-looking doll. It's an image which fighter jocks have worked hard to earn.

One of the qualities for which the astronauts were selected, supposedly, was self-discipline. Not everyone was seduced by the lifestyle, but you can believe that all were tempted. I'm only surprised more didn't succumb to the easy times and promising nights.

As a group, I suppose our moral code was about what the national average would indicate—if there is a national average. It is even possible, under the circumstances, that our behavior was better than the gossip and suspicion implied.

Sure, there were astronaut groupies who were not much different from those who tracked rock groups and baseball players. They called and wrote notes and stood ready on a moment's notice to whisk us off to paradise. Some even restricted themselves to only the well-known, a most devastating kind of discrimination. The "specialists" might not recognize us if we fell in their laps, until the instant we returned from a flight and qualified for their target list. There was one young astro who hungered after a long-legged beauty at the Cape, a good "friend" of the big boys. She wouldn't give him the time of day. However, the very night of his return from his first space mission she literally fell into his bed. She was one of several who managed to collect at least six or eight of our gang. With some it was clearly a sport, and they were proud of their scorecards.

One woman supposedly presented herself at the offices of *Life* magazine, in New York, offering to sell a story that she obviously thought was a blockbuster. It was her proud boast—which she planned to document in vivid detail—that she had taken to bed every one of the nation's astronauts. She had batted a thousand. *Life* doubted the authenticity of her claim and, besides, she hardly squared with the image the magazine had promoted of the astronauts as America's Galahads.

Later, when an unofficial survey was taken around the office, there wasn't a single astronaut who even recognized the lady's name. Believe me, we tried.

Miss Tell-It-All wasn't the first to seek publicity for notoriety that may not have been deserved. Every so often we'd pick up a gossip column and read that one of us had a little action going with this or that celebrity. For the next few weeks the wives would be on the hot line comparing notes. After a while they learned that the best measure of a story's lack of substance was the very fact that it had appeared in print. No astronaut needed that kind of exposure, but the woman involved always did.

One item in the "you'd-never-guess-who-has-been-seen-together-lately" category involved astronaut Charles (Pete) Conrad and singer Keely Smith, Louis Prima's ex. I first heard about it the way I usually got such information—through my wife. Naturally, she wanted to know if there was any truth to it. It sounded absurd and I told her so, but I did take the time to ask Pete how such a story could get circulated. "It's easier than you think," he said, with a pained grin.

The whole thing began a week or two before. Pete and some friends were having dinner in a Los Angeles restaurant. During the evening a friend at another table had brought Keely Smith over for an introduction. They chatted for about five minutes and, as she was leaving invited Pete to stop by at a birthday party she was giving for one of her children the following Sunday: "The kids would love to meet you." Pete made a polite but nonbinding response—"Well, I'll try"—and finished his dinner. Their paths never crossed again. End of story.

Two days after that story had been circulated I telephoned Lo from St. Louis, and she eagerly gave me the latest bulletin. "Betty called," she said, giggling slightly. "She heard we were splitting up because you've been dating Keely Smith!"

That's where I came in.

A more persistent story was one that linked Neil Armstrong and Connie Stevens, the former wife of Eddie Fisher. The story popped up several times over a period of months and always in the region where her act was booked. It was completely out of character for Neil, both in terms of his own behavior and his sensitivity to public opinion. This isn't to portray us as some kind of underdog, but in these instances the astronaut had little to gain and virtually everything to lose. For the entertainer, the major consideration was to get her name spelled right.

The bachelor astro was only slightly better off than his married compatriots. His personal life might be able to stand the heat, but he still had the space agency looking over his shoulder. Gossip of *any* kind made NASA *very* nervous.

When it became public knowledge that one of the first scientist-astronauts, Dr. Duane Graveline, was having domestic troubles, NASA didn't hesitate to get into the act. Not long after his wife filed for divorce, the agency quickly asked for and received his resignation. That was as dramatic an example to the rest of us as a neck attached to a swinging rope was to horse thieves in the Old West.

There were other examples of how NASA took a personal interest in the private affairs of the astronauts. When John Young became involved in a long romance, he was summoned, or so we heard, for a fatherly chat with Dr. Gilruth, the head of the Manned Spacecraft Center. No one knows what was said behind Dr. Gilruth's door, but the rest of us watched closely to see if he would remain in the flight rotation. He did. (And he married the woman.)

At times we welcomed the help and influence of Big Brother. On one occasion Wally Schirra suddenly found himself with a second wife he knew nothing about. The lady had signed herself in at a Cocoa Beach motel—partly owned by the seven original astronauts—as Mrs. Wally Schirra. She began spending money, cashing checks, and advising everyone within earshot that she was the second and genuine Mrs. Schirra. People were beginning to wonder, and Wally finally turned the matter over to NASA security. They exposed the woman and sent her packing. No one knew where she had come from or where she went.

Over the years, the astronauts also acquired a semistraight following that mostly parked in the motel lobby and looked at the scenery. We referred to them as "the little old ladies," although in truth they ranged from teenagers to the Grandma Moses type. Most of them were tourists, and they were content just to be where the astros hung out. To avoid a big autograph scene, we developed a technique of hurrying into the lobby, registering on the run and heading for the room without looking left or right. Looking back, I'm sorry now we didn't take a little more time to oblige them. What those ladies had in mind was motherly love, and you don't have to play games to earn it.

Part of the way we lived was a response to the pressure, the hours, the womanless world in which we worked and part was biological. But much of it was purely entertainment and did not always lead to the boudoir. In fact, we often fumbled away opportunities—opportunities we had angled to get.

Once, I was to speak at a UCLA alumni dinner in Los Angeles, at a swank department store where NASA had placed on display some of the Apollo 11 moon rocks. (Coincidentally, some of them were stolen that night, the first such theft of samples brought back from another heavenly body.) A striking blonde, whom I judged to be about twenty-five, caught my eye during a fashion show that featured some of the local socialites. I was at the head table, sitting between the wife of the UCLA chancellor and the woman who was the evening's program chairman. When the fashion show ended I was surprised and flattered to see the blonde I had singled out walk straight toward me. She held out a program, smiled, and said, "Could I have your autograph for my children?"

This was one of those times when it was anything but annoying to sign an autograph. Returning the smile I said, "My pleasure. What are their names?"

She began rattling them off so fast I did a double take.

"Just a second. How many kids do you have?"

"Five."

I was stunned. "Well, how old are they?" I wasn't just being curious. Before writing down, "Eat your spinach" or "Study hard," it helped to know if the kids were five or fifteen.

"The oldest is seventeen," she answered.

"I don't believe it," I said. "How could you have five kids from seventeen on down?"

"Oh, it was easy," she responded. "I just screwed a lot."

I ducked my head and started writing as fast as I could, not daring to glance at the chancellor's wife. I muttered something like, "Well, I hope you stay in practice," and handed back her program. She smiled and walked off. That night it wasn't easy to pay attention to the program.

The story that probably best illustrates our collective attitude produced an enduring line that could almost have been our motto. Conrad, Gordon, and Bean (the Apollo 12 crew) were friends long before they were teamed up as astronauts, and they were a pretty unflappable combination. They seemed to enjoy themselves regardless of the occasion. After making the second lunar landing, they were guests at a press luncheon in Oklahoma City. Through some private signal, they had called each other's attention to an attractive female reporter in the audience. There was a lot of smiling back and forth, and when Al finally caught her eye he threw her a wink. She caught a lot more than Al was pitching, because when the conference was over and he walked through the lobby, there she sat, waiting for him.

Al paused only briefly. "Sorry honey," he said. "Just testing."

Testing or not, an astronaut generally expects a woman to accept if he asks her out. And she generally does.

We never lacked for vicarious thrills, and some that were not so vicarious, but we tried not to confuse romance with matters of life and death. Jobs that involve any degree of risk seem particularly appealing to men who believe in their own machismo. Most of us like to think that we have a little Don Juan in our blood, if only we had the time. The other wives teased Dick Gordon about his sex appeal, his animal magnetism. But I'm not sure Dick appreciated it when someone dubbed him "the animal."

At the other end of the spectrum were guys so straight that you didn't know whether to admire them or have them stuffed and shipped to the Smithsonian Institution as the last of a vanishing species. The likes of Frank Borman, Stu Roosa, Bill Anders, and the late Ed White represented a threat, and a challenge, to the lovers in the group. They were a continuing reminder that no law of nature said you had to be a swinger. Some of the wilder ones would have felt more secure had they been able to bend some of the straight arrows by offering them favored phone numbers, but to my knowledge they never succeeded. That was alright, too. Regardless of our preferences, we didn't preach or make moral judgments.

Any group of seventy-three reasonably normal males placed under a microscope could produce results not unlike ours. Some fell in love; some broke up their homes; some ran up the score; some kept the faith. Few could resist the battle of wits.

One well-known astronaut was warned by a NASA official that his reputation was getting out of hand. He was told to be more discreet. A few nights later, he drove fifty miles south of the Cape to take a lady friend to dinner. When they sat down and looked across the room, the first face he recognized was the NASA big shot, quietly dining with his own beautiful companion.

Another time, several of us had the somber job of cleaning up the personal effects of an astronaut who had died. As we went through his desk and files and closet, his closest friend let out a wounded yell and held up a small address book. "That son-of-a-gun was messing with my girl."

With all the competition that went on and all the potential that existed for mischief, the astronaut record of seven divorces in sixteen years isn't very scandalous. The seams of several other marriages underwent great strain. The wonder is that more didn't tear apart.

Of the forty married astronauts who flew in space, seven have gotten divorced—roughly 15 percent. Six of those divorces occurred within a two-year period, leading some critics to suspect we were coming back from space with some kind of social disease or a postorbital melancholy.

Time, and the fickle interest of the public, have made it possible for an astronaut to get a divorce today and not make news. But it wasn't always so. In fact, the first astro divorce—which took place in 1969, after ten years of manned space flight—was a landmark case for us.

Although he wasn't the first to consider it, Donn Eisele was the first to divorce his wife. She had joined the program with him, had enjoyed the good and the bad, and she had been a part of *Life*'s happy-home-life image. Donn's friends didn't judge the merits of his action or try to decide who was at fault. We saw it as a test case to determine whether an astronaut was free to get a divorce without endangering his career.

At the heart of this cause célèbre was a lovely young woman whom Donn had met early in 1968 at a Cocoa Beach party. This was the year we practically lived at the Cape and were regularly involved in the social scene. Susie, who was single, was friendly with the local married set that so often played host to the astronauts. She had been to several of the parties before Donn met her, and some of us knew her as a pretty and intelligent young woman. At the time Donn and Susie met, he was experiencing deep personal problems at home. Harriet Eisele wasn't known for her understanding about the travel and other demands of Donn's job and she carried the added burden of loving and caring for a retarded child with leukemia.

One night the Eiseles entertained the Apollo 7 crew, and most of our wives attended. Jack Swigert, a bachelor, was there with a new girlfriend (we seldom saw Jack with the same woman twice). As the evening wore on and the drinks flowed, our wives began commiserating about their sorry lot and how particularly tough it had been with their husbands traveling so much. Jack and I were standing to one side listening. Finally, Jack had heard all he could tolerate. He got on his own soapbox and took a position for which most of us privately applauded him. Most of his remarks were directed right at Harriet Eisele.

"You wives are foolish as hell," he said, "if you take this attitude every time your husband comes home. Yeah, we've been off all week in the land of milk and honey and beautiful secretaries. But we're also working our tails off. Your husbands look forward to coming home for a little loving relaxation, not a lot of nagging. You should make them feel great when they get back to town, not like they're taking a dose of medicine. If you don't, some weekend when he's faced with a choice of staying where he is or flying home, he isn't going to make the right decision."

I looked at Harriet Eisele. She seemed amused, certain that Jack had overdramatized the situation. The other wives weren't so sure.

I thought of Jack's brief lecture in the days when Donn began to lose his enthusiasm for rushing out of the simulator on Friday at the Cape and making an early takeoff for Houston. He began to volunteer for more than his share of late Friday and weekend duty. We knew what was happening: he was getting serious about a relationship we had all assumed was a casual fling. From time to time Wally or Tom Stafford or I—in fact, all of us—offered him some unsolicited advice. We suggested he go slow.

I don't believe Donn resented our interest. He knew we thought we were looking out for him. At that point, no one suspected that he was headed for a divorce and a new marriage. His free-time life at the Cape did not interfere with his job until the last two weeks before the flight, so we really couldn't butt in too much. He had been around long enough to feel that his behavior was more typical than not.

As launch day approaches, the flight crew attempts to block out the outside world. That effort reaches a peak the final week. As October 11 and the lift-off for Apollo 7 approached, we weren't getting home at all. Wally and I became concerned that Donn was feeling too much of the prelaunch pressure or Susie or both. That last week he zipped into Cocoa Beach at every opportunity; the last two days he wasn't even available for several scheduled activities. We were worried, not because he was wrapped up in a girlfriend, but because at that point his mind should have been on a mission that was only hours away. We had obviously underestimated just how serious his affair was.

Following the flight of Apollo 7, Donn was assigned to the backup crew for Apollo 10, then training at the Cape. The romance continued to flourish. When word got out that the Eiseles were getting a divorce, the first reaction around the Astronaut Office was that Donn had fallen overboard. He had violated the first rule of philandering: "It's okay to fool around but don't fall in love."

If the story was only of passing interest to the public, it really had our attention in the office. An astronaut divorce, especially the first one, was bound to attract the notice of the national media. I suspect that Donn already had a timetable in mind for his departure from the space program when he arranged for his divorce. Susie wouldn't be just a new wife, moving into the tight little space colony. She was a woman who had traveled in the same social circles as the guys when they were at the Cape. Susie knew us living the bachelor life of men away from home, and it just plain scared hell out of some, the idea of her getting chummy with our wives.

Not that they really needed to worry. Under the circumstances, when their paths crossed, most of the wives treated Susie as though she were Typhoid Mary. In their view the new woman on the block was married to the "bum" who walked away from his responsibilities—including a retarded child. They could identify all too easily with Harriet.

The astronaut wives held a monthly coffee and the move of Susie Eisele to Houston was probably the most exciting topic of conversation to reach their tables in years. Lo, JoAnn Carr, and Beth Williams, C. C.'s widow, were in Susie's corner and felt she deserved to be received with courtesy in spite of any negative feelings toward Donn. But there was no way the majority of them were ever going to accept Donn's second wife.

Susie Eisele turned out to be a fighter. After she recognized the frosty reception in store for her, she quietly passed the word to cool it—the old "people in glass houses" gambit. Many of us, whose Cocoa Beach behavior was not exactly spotless, implored our wives to thaw things out a little. But a few of the diehard wives went so far as to schedule parties when Donn was out of town, so they could avoid inviting Susie and having to speak with "That Woman."

I don't believe his decision to leave NASA was a very emotional one for Donn. His performance the preceding six months had pretty well alienated the powers in the astro office. His career was definitely not upward-mobile. Yet we were all interested in the fallout that was sure to come. There were several astronauts, including Scott Carpenter and Gordon Cooper, who had been holding failing marriages together, hoping for another flight. Even those of us with solid marriages waited for the aftershock. Would it bring on a surge of divorce filings? Would life go on? Would NASA try to put as much distance as possible between itself and any culprit who besmirched the Boy Scout, white hat, All-American boy image which they had so carefully nurtured and protected?

Donn was an unlikely candidate to challenge the system. He was not one of the in group, and most of us in the office could identify with his position in the pecking order. We watched fascinated, not by the personal drama, not with deep concern over the breakup of a family, but with curiosity about the effect on Donn and his career. And ultimately, the implications for the rest of us.

Fair or not, when Donn married Susie it was inevitable that he would resign from the Astronaut Office. He was encouraged to do so by some of the fellows, who were uneasy about how much Susie knew and whether she would respect their confidences once introduced socially to their wives. Tom Stafford was the head of the Astronaut Office then and, until that time, one of Donn's good friends. In that dual role, the fraternal efforts to keep family skeletons in the closet at the Cape fell on Tom's shoulders, and he did his best to influence Donn's thinking. Donn knew the score and, I suspect, felt he was keeping faith with his friends when he made the decision to resign. That decision was hastened by the systematic exclusion and a growing disenchantment with his job. It was greeted with relief from most quarters. He accepted a position with NASA at Langley, Virginia, and a short time later left there to join the Peace Corps.

Donn Eisele didn't exactly become a martyr to his fellow astronauts, but at least he had broken the ice for those who felt trapped in dead-end marriages. The prevailing opinion in the office had been that our position as public figures and NASA's obsession with the astronaut image demanded a willingness to tolerate domestic pressures. Now it was different. No one envied Donn his fate, but at least the earth hadn't just swallowed him up. The agency hadn't gotten directly involved in any obvious way. No one had forced him out. He might even have gotten away with it if he hadn't married an old girlfriend and settled her in Houston. This rationale looked fairly tempting to those who wanted out of a collapsed marriage.

Think what this meant. Holding things together was part of the price each was willing to pay for that ticket into orbit. Each had been willing to subjugate his own personal feelings, his desires, his home life, good or bad, to keep the gods of space happy. Now someone had crossed the line. It was emancipation day.

In sixteen years, forty-one married astronauts have gone into space at least once, and it was from this experienced group that the divorces came. Among the thirty who have not flown, only one, Don Holmquest, cut the cord from his original wife. It had no apparent effect on his career, which was already on the skids.

For a time the favorite question of the news media was: Did space flights somehow change us, make us restless, looking for new directions, new hopes, new wives? The answer is really quite simple. It was a case of men who were used to taking charge, finally returning to the helm of their own lives, and making decisions they had postponed in order to protect their careers. Each astronaut took the step at the end of his flying career.

Scott Carpenter, Gordon Cooper, John Young, Donn Eisele, Ed Mitchell, Al Worden, and Buzz Aldrin were the astronauts whose marriages broke up. All seven had made at least one flight, and none would make another. Curiously, after Eisele made the big move none of the later divorces caused much of a ripple around the office or the home front. The last I heard, Donn and Susie were back home in Virginia and happy. It isn't the kind of thing they give medals for, but Donn did make it easier for the others to pursue normal alternatives to their married lives.

Most of the marriages were strong enough to withstand more than the usual load of bumps and shakes. Those who strayed were prudent enough not to let it affect their families—with the exception of Donn and John Young, who divorced his first wife to marry a woman he had loved for years. The others who did obtain divorces followed the usual pattern of eventually remarrying. Does anyone really find anything sensational in these figures? Or is it a fair cross-section of human behavior, given the odyssey we all shared?

The most publicized astronaut infidelity was a crisis in Buzz Aldrin's marriage. He told his wife about the other woman, and in his book, *Return to Earth*, he told the world. My own reaction was that Buzz had shown some guts, but little headwork. Buzz thought he was in love. His wife demonstrated patience, and they worked it out *on that occasion* (Buzz later divorced and remarried). What else is there to judge?

Society has come a long way since the days when Elizabeth Barrett Browning wrote, "How do I love thee? Let me count the ways," to a time when Masters and Johnson did the counting. My only other reaction to Aldrin's published confession was surprise that so many people seemed unprepared to learn that an astronaut had human appetites. Somehow, I don't believe it weakens the great technological achievements won in space to know they were accomplished by living, breathing, emotional, human beings, and not by cold, programmed automatons.

Soon after my own selection I came to realize that being an astronaut would elevate, in some vague and unbiological way, my own sex appeal and social status. The first indication came the day after I was selected. The headline in the next edition of the Santa Monica paper read, "RAND SCIENTIST SELECTED AS ASTRONAUT." The next morning I had a phone call from a woman I hadn't seen since high school, fourteen years before. In the intervening years we had each married and settled in the same city, yet had never bothered to contact each other. Now, it was as if the ninety-seven-pound weakling had turned into Charles Atlas overnight. Carol came by the office; we had lunch and then strolled along the beach reminiscing about our high school days, when we had dated. I could almost feel the sparks flying, even though I was the same craggy-faced, studious fellow who one day earlier would not have received a second glance. Nothing had changed, except for one story in the evening newspaper. I went back to my office that afternoon feeling slightly heroic. When I told Lo about my lunch date, she too was amused.

What makes this subject so delicate is the fact that nearly all of the astronauts were married, although some were married in a very limited geographical sense. During this period, to have been an astronaut *and* a bachelor, well, that was as close to paradise as you could get. Just imagine, to have the test pilot swagger, an ample salary (with little overhead), a jet airplane to hop around in, and the very highest professional respect. Prize bait in the social fish bowl.

Among the legitimate astro bachelors—C. C. Williams, Jack Schmitt, Ken Mattingly, and Jack Swigert—only C. C. and Swigert seemed to capitalize on the great potential of the situation. When C. C. got married before the end of his first year at NASA, it was up to Swigert to carry the flag.

Jack's coast-to-coast girl chasing was in true archetypical bachelor style. For short periods Jack would have someone special in his life, someone he thought a little more of than the others, someone to act as hostess at his annual wine-tasting party, someone our wives might recognize from one social function to the next. Jack was a cross between the swinging bachelor and the confirmed bachelor. He was a fastidious housekeeper, who complained little about living alone, and always seemed informed as to the best buys on meat and canned goods.

Lo and Jack had always gotten along famously, and on occasion she would fix him up with someone we knew. Once Lo arranged for Jack and my younger sister Charlotte and the two of us to go out for dinner. My only concern was whether Charlotte could take care of herself. After all, Jack went through women like Sherman went through Georgia. But no cause to worry. On the appointed evening Jack resembled the Tony Randall role in "The Odd Couple" more than the *Playboy* image of a bachelor. Rather than exchanging teasing banter with his date, Jack and Lo spent the half-hour drive to the restaurant discussing the comparative costs at local supermarket chains.

It is no betrayal of fraternal trust to say that Jack carried on a relentless investigation of the he–she relationship. He covered the full spectrum of female companionship, from White House file clerks to Hollywood starlets. Jack was the kind of fellow who could work all day in Houston or at the Cape and casually make a date for the same night in Los Angeles—and keep it. In his job he was dedicated; in his private hours he was indestructible. It wasn't that Jack scored every time—if he had, his body would now be in a jar at the Harvard Medical Laboratory—but no one ever devoted more time or energy to the chase.

In some respects, Jack's MO—as they say in police work—was the envy of his peers. He had probably collected the longest list of names and telephone numbers of anyone I have ever known. Whereas many of the guys simply indexed their black books by state, Jack's was truly a living file. One imagined a steady stream of names moving from the category of hot prospects to active players to recently retired. And sometimes he would recycle them as oldies but goodies.

Anyone who traveled with Jack was witness to part of his technique at refueling stops all over the country. Typically, a T-38 pilot would have just enough time to file a flight plan and grab a cup of coffee or a candy bar while the ground crew refueled his aircraft. Whenever Jack pulled into the line, he was in just as big a hurry as the rest of us, but his rush was to the nearest telephone. It didn't seem to matter whether we stopped in Atlanta or Point Desolate, Maine. There was no place beyond the reach of Jack's little black book. If he had the time, he would check in with every girl in the near vicinity, those he had dated and those he was "auditioning." It was a quick, friendly "just-passing-through-and-thought-I'd-say-Hi," but it went over big with the callee. We rode Jack without mercy about it, but he seldom missed an opportunity to make a call.

To his credit, Jack never neglected his job, even though he was constantly trolling for a new catch. The impressive thing about Jack was his ability to operate coast-to-coast, thanks to the instant mobility provided by the T-38. I'll never forget one of Jack's four-day grand tours in the summer of 1968. He was working at the Cape, helping us prepare for the Apollo 7 mission. We had a long holiday coming up— the Fourth of July weekend—and as we talked one day Jack proudly related the man-killing schedule he had laid out for himself. Of course, it was planned as a triumphant march across the continent, from the mountains to the prairies to the ocean white with foam. On Wednesday night he would fly to Miami to audition a new prospect he had met down in stewardess heaven. On Thursday it was back to Cocoa Beach for a date, to Atlanta for a hot one Friday night; into Denver, the old home town, for a Saturday night special; and then the big windup Sunday in Sacramento, California.

When Wally Schirra wondered when he had found the opportunity to line up a girl in Sacramento, Jack's answer was right in character. "Remember the press conference we held at the Aerojet General plant last year?" he asked. We all nodded yes. "Well," he said, grinning, "you remember the nice-looking female reporter who followed us around on the tour?"

We all nodded and understood. Jack had kept her on the back burner for ten months, and Sunday night was *the* night. We could hardly wait till Monday to see if he was still alive.

Jack lived through many such trips with no ill effects. He was single-minded and methodical in affairs of the heart, but his success was in no way comparable to his efforts.

Still, among the bachelor astros, he was clearly the team leader. C. C. Williams was active, without being obvious. Ken Mattingly gave the impression that being an astronaut was part of a larger pattern. He was self-disciplined and led an almost ascetic personal life. Then there was Harrison (Jack) Schmitt, who seemed almost wary of the astronaut reputation and the women it attracted. Schmitt was short, dark, handsome, and sometimes alarmed by the more aggressive, liberated females. Office gossip had it that more than one good-looker had been told by Schmitt, "Look, I'm not that kind of a guy. I don't just tumble on the first date. It means more to me than that."

Swigert was the predatory male in the classic sense. And though we had few bachelors, he wasn't without competition. It came from the few really swinging married guys who, on occasion, worked the same territory as Jack. This was a source of continuing irritation to Jack, who felt his flanks constantly threatened. Whenever he learned that a married buddy was dating one of his girls, Jack would lecture her solemnly on the moral implications of her conduct. "Look," he'd say, "you shouldn't be dating him. He's married. He has kids. You should be dating someone single. Someone without obligations. Someone like me." How much of an impression this appeal made often depended on whether she was herself married.

I don't recall the competition ever turning surly. No punches were ever thrown in defense of any lady's honor, real or imagined, and in a way that's odd when you consider how much sharing was going on. The married guys were amused by the earnestness of Jack Swigert. They were in it for the thrill of the sport, as one of them made clear one day when he made a pass at a secretary in the Astronaut Office.

In the friendliest of gestures, he patted her on the fanny and said, "Say, when are you and I going out to dinner?"

She looked at him innocently. "I couldn't do that," she said. "You're married."

"That's okay," he assured her. "I'm not looking for a wife."

If we begin to sound like so many traveling salesmen, that may not be a bad analogy. We were selling ideas, to engineers or managers or to the public. We were on the road half the time. Our territory was all fifty states and points beyond. We had a shine on our shoes and a smile on our faces. We had one advantage no traveling salesman ever enjoyed: the T-38, our favorite fringe benefit. The Beautiful People with their private jets had nothing on us.

In pilot jargon, the T-38 was a very sexy aircraft. It was hard to look ordinary climbing out of the cockpit of that airborne white stallion, decked in full flight regalia. The T-38 was to many of us the symbol of the astronaut scene, embodying everything essential to the definition of glamor. And rare was the lady who was immune to a personal guided tour through the cockpit of this sweet airplane. Such was its impact that it eventually gained the unofficial nickname "the golden leg spreader."

I can see where we developed a kind of split personality. There was the astronaut image; the pursuit of a life-style that in other places or other times would never have been accessible to us. Then there was the side that shrank from the image and the fanfare, when we reminded ourselves that at the bottom we could be dull men, sent to do a technician's job.

If we revved our engines at times it was partly to change the pace, partly out of restlessness—our way of burning off excess fuel. More often than not we sought the company of trusted friends, who sensed our moods, knew our rhythms, could keep a confidence, and didn't go bananas if they discovered us in a pub having a drink with one of the local beauties.

That's what we liked best about Cocoa Beach. It encouraged a casualness, an almost barefooted earthiness, that is common to small Florida towns. But there was action if you wanted it. An astronaut could run wild within the limits of his own endurance and the concern he had for his reputation. (Some had very little.) I was standing at the check-in desk of the Holiday Inn one day, when the rather plain clerk behind the counter nodded across the room and said, "Isn't that so-and-so?" referring to one of the more popular astronauts. "Yes, I think it is," I said. She shrugged. "I bet I'm the only woman in Cocoa Beach that hasn't been to bed with him." I went away wondering if she was bragging or just feeling neglected.

It wasn't all Ban-Lon shirts. happy tours, and souped-up sports cars. Most of us developed friendships with local couples who made us feel comfortable and provided a kind of sanctuary when we needed it. We could let our hair down with the Bishops or the Johnsons (if our hair had been long enough to let down). In their company we could get snockered if we had a notion to and not worry about the local gossip columns describing how we disgraced ourselves.

They were people who made no demands and asked few questions. They were always ready to throw a party or babysit with our kids or roast wieners over a beach fire while someone strummed a guitar. Gayle Harris always had a reserve of attractive friends, some married and some single, who were part of the scene. The price of this friendship was to be gossiped about in the parlors of what passed for Cocoa Beach society. There were enough stories floating around about the astronauts to indict any woman seen in the same room with them. Generally, these activities were about as sensual as the Andrew Sisters and the Stage Door Canteen.

We dubbed Gayle and her court "The Vestal Virgins." They would flirt and dance and have a good time, but they'd depart early for the home fires while we dragged ourselves back to an empty room at the

Holiday Inn or tore up the highway back to the crew quarters at the Cape.

We also had great times at the home of Al Bishop, a space contractor field represenatative during the heyday of the space effort. I'll never forget one blast of a party, the night before a launch, that lasted almost to lift-off the next morning. The highlight of the evening was when Donn Eisele decided to swim his way back to the Holiday Inn, three or four miles away. Donn dove into the canal behind the house while the guests looked on, their moods ranging from disbelief to hilarity. He splashed out of sight before it occurred to anyone that in his inebriated condition he could drown himself. A search party spent the next hour and a half patrolling the water's edge. Finally Donn was discovered fast asleep in someone's backyard.

The crew quarters NASA supplied at the Cape served as a combination dormitory and locker room. Until the preflight quarantine became mandatory after the measles scare of Apollo 13, we lived there more or less at our own convenience. They were handy during late night testing or long schedules, or just to take a breather from the social vices. The crew quarters had just two knocks against it: if we stayed there we lost our twenty-five-dollars-a-day per diem, and it was twenty miles away from the action on the beach.

NASA management, of course, looked upon the crew quarters not only as an amenity, but as a means of exercising control and cooling down preflight shenanigans. Even the location and design contributed to that end. They were located on the third floor of a building devoted to engineering, testing, and office space, with people coming and going at all hours. But through skill, cunning, and a high degree of motivation, the astronauts still managed to bring off an occasional party. There was always a hard core who felt that a failure to do so would constitute a waste of available resources. The traffic in the building, and the presence close to launch time of as many as fifteen astronauts, did little to discourage such efforts. For months afterward, the boys were still talking about a social gathering that must have taken place shortly before one of the last Gemini launches. Late into the night, or early into the morning, they must have gone into the sauna en masse to bake out.

No one might ever have found out except for two clues discovered by the housekeeper the next day: an empty wine bottle and a bra. Of course, it was possible that someone planted the evidence for a good laugh or to cast suspicion on the crew in training. Incidents like that didn't endear us to management, needless to say, but they made life a lot more agreeable.

Let's face it, when a person decided to exploit it, the sexual fallout could be a significant fringe benefit of being an astronaut. While the temptations were great to measure up to this picture of the glamorous fighter pilot, man on the move, we did not all succumb. Some stayed clean because of their moral character but many others for practical reasons. Traveling so much at night and working during the day could put a big dent in the number of opportunities to make whoopee. It was tough enough trying to keep up with your homework. There were times when one's resolve would be sorely tested by the situations in which he found himself. We didn't have to have a roving eye to find the action. A man not usually inclined to play the game could still be a pushover for a woman who fell into his arms. It still takes two to tango, and a lot of women wanted to dance with an astronaut. They acted like sleeping with him was something special indeed.

Keep in mind it was a small club; most of us knew what the other guy had going. Some of the guys developed their own unique reputation. One of the older heads was known for the great quantities of tail he collected, while another concentrated on quality with only a small sacrifice in numbers. A third gained a reputation as a man who would screw a snake.

Against that background of overachievers there's no doubt about it; after some astros left the program their wives sex life improved while theirs took one hell of a beating.

I should note that I have not attempted to justify the way we lived and how we behaved. Whatever the space program achieved in the 1960s, whatever it meant to the American spirit, was made possible by men who responded superbly under extraordinary conditions. The American public wanted heroes, and heroes they got. I just don't believe it dulls those achievements to know that the men who became the symbols of them were otherwise rather normal.

12

Footprints on the Moon

O N J U L Y 2 0 , 1 9 6 9 , at the moment their boots made contact with the lunar dust, Neil Armstrong and Buzz Aldrin literally stood on a new plateau for all mankind. Though subsequent landings would each be more difficult technically, all would come to be regarded as a kind of encore—a reworking of the path Apollo 11 had blazed before them.

Yet somewhere between Neil Armstrong's small step and mankind's giant leap, came the strides taken by Apollos 7, 8, 9, and 10, enabling America to go that last historic mile. Of the twelve crewmen involved in those first four Apollo missions, only three (Scott, Young, and Cernan) would return to walk on the lunar surface. But all of them—all the men of Apollo—would leave their footprints on the moon.

At the height of our national rapture over the landing, Olin (Tiger) Teague, the portly congressman from Bryan, Texas, had an idea. Teague called Deke Slayton and suggested that he, Teague, should get the ball rolling to recommend the Apollo 11 crew for the Congressional Medal of Honor.

Deke gulped. The crew of Apollo 11 had already received the Medal of Freedom, our highest civilian award, personally from President Nixon. Honors were being showered on them at a rate which may have surpassed any other single achievement of man. The general sentiment was that all were deserved.

Slayton listened quietly as Congressman Teague pointed out that the

precedent had been established forty years before when Charles Lindbergh became the first man to fly the Atlantic.

It must have been a delicate moment for Deke, whose pride in the
success of Apollo 11 was equal to anyone's. But he would oppose any
move to single out one astronaut or one crew for the nation's highest
honor, and he said so to Olin Teague as firmly and gracefully as possible. He maintained that such an act would be a slight to the rest of
the team who had, in effect, run interference for Neil and Buzz and
Mike Collins.

There was another consideration, Deke's own awareness, shared by
a few others, that the roles of Armstrong, Aldrin, and Collins in this
historic event had been determined largely by chance—by the luck of
the draw. Years later, as the Apollo program was ending, Deke confided to me that he believed he had prevented the Apollo 11 crew from
receiving the Congressional Medal of Honor. It wasn't hard to read
him. Deke had wrestled with his conscience at the time, and he was
still a little troubled by it.

In no way do I intend to minimize the achievements of Apollo 11, a
flight that was a premier event in the long saga of man. But unlike
Lindbergh and Columbus, who struck out on their own, without promise of success, Apollo 11 would be selected in an impersonal way to do
something immortal.

Sir Edmund Hillary, with Norgay Tenzing, made the first successful
assault on Mount Everest by walking in the footsteps of their predeces-
sors. Many with Hillary worked their way to the upper camps where it
was decided who would make the final successful push the last short
distance to the summit. Armstrong and Aldrin were the Hillary and
Tenzing of the Apollo program. The landings which followed Eleven
all traveled the same path in space. They used the same building
blocks, though each made a slightly different excursion at the end. The
first to put together the building blocks of the four preceding missions
was Apollo 11.

Who would have thought that the fifth giant step would put a man
on the moon? Certainly not the astronauts. We were too smart. Oh,
yes, we were aware of the plans which, if successful (and it was a
mighty big *if* in 1967 when the mission sequence was formulated),
would place Apollo 11 in a position to attempt the first landing. We
just never expected success to come that smoothly.

I would characterize those five giant steps in the following way:

Apollo 7. Successful, but it left a bad taste—confirmed our spacecraft
        was safe to fly.
Apollo 8. Spectacular but superfluous. There was even time to say a
        prayer at Christmas.

Apollo 9. Unsung. Will the Lunar Module work? That time the prayers
were for Rusty Schweickart to get well enough, soon enough, so
the mission would not have to be repeated.
Apollo 10. A dress rehearsal. We spat in the eye of the man in the
moon.
Apollo 11. Big casino.

From October 1968 to July 1969 we were caught up in an inexorable
string of successes, almost as if it were beyond our control. We had no
time to figure out what we were doing right, only time to keep on
doing it. With each lift-off we were only two months away from the
next.

When Apollo 7 lifted off on October 11, 1968, on the most ambitious
first flight ever attempted, it all boiled down to one objective: prove
that the Command and Service Modules comprised a safe vehicle
for man's travel beyond the earth's atmosphere. The task wasn't only
technical. We also had to address any psychological barriers that still
remained. It was essential that confidence be restored in a vehicle born
to problems and disaster. We had remade the Apollo spacecraft incor-
porating every operational and safety modification for which a case
could be made. The Seven crew believed in it. All that remained was to
take her out of the barn and fly.

Our success, of course, was crucial to the established mission sequence. The space agency had quietly decided a month before our own
lift-off that Apollo 8 would travel to the moon—*unless* we turned up
something unexpected. From that point on the Apollo 8 crew would be
hanging from the light fixtures until they heard us say, "No sweat. It's a
great bird and you can count on it." And that's precisely what we
reported.

Recall that Seven was the second Apollo spacecraft. The first had
burned on the pad. Seven accomplished its job by eliminating any
lingering apprehensions, which permitted Borman, Lovell, and Anders
to set sail for the moon with confidence.

For the first time, on Apollo 8, men in space would be not just thirty
minutes away from terra firma, but as long as four days away from
their home planet. To achieve that milestone they were prepared to
accept higher risks than the NASA managers who had to give the final
go-ahead.

Apollo 8 provided a tremendous emotional boost to the space program for insiders as well as spectators. It proved we could go the lunar
distance and, for the first time, the country *knew* we were going to
beat the Russians.

In 1968 NASA still believed, or said it believed, that the Soviet
Union was battling us neck and neck to put the first man on the moon.

The Russians had upstaged us so often that the agency was scared silly they would attempt something spectacular prior to our first lunar landing. We had a way to head that off and give the Russians a show. The Lunar Module wouldn't be ready to fly for several months, but we could send men *around* the moon with what we had.

There was little more that could be accomplished on a circumlunar flight than had been planned for the original high apogee earth orbital mission. The only things new on board were some of the programs in the Apollo computer. The one factor added to the Apollo 8 mission was the distance from the earth—and the potential for a long, slow return in the event of an aborted mission. Everyone was hell-bent for going to the moon, including the Apollo 8 crew. The go-ahead was given.

In its journey, Apollo 8 would experience two sights never before seen by man: the dark side of the moon and the planet earth in its full global splendor. How I envied the crew! In those moments when I had allowed myself to dream of a private experience to carry with me from my years at NASA, it was always of the first view of our planet suspended in its pristine beauty against the black velvet fabric of space.

The first landing on the moon was simply beyond my imagination. The flight sequence was against me, and I dared not covet an honor that seemed beyond ambition. Besides, I asked myself, why should fate place its fickle finger on me? I would assume that most of us went through a similar monologue. It was a way of preparing oneself for the disappointment if the plum fell to someone else—as it did. *Sic transit gloria mundi.*

The big guessing game had been accelerating ever since the office summit meeting of April 1967, when Deke called in the prime and backup crews for the first three Apollo missions. It was then that he announced, "The eighteen of you are going to fly the missions that get us to the moon, including the first lunar landing."

We proceeded to lay out two plans; one optimistic and the other extremely optimistic. The first plan called for a landing on the fifth manned mission without a prior close approach to the moon's surface. The other called for a landing on the fourth mission, likewise without benefit of a close approach. (See chart.)

At the time none of us knew on which mission a lunar landing could be safely attempted or how many unsuccessful attempts might be required. Deke wanted an experienced nucleus from which to make crew assignments. Those crews would then rotate their way toward the top of the pyramid. We had the feeling we were reaching the climax of a grand competition. Actually, we had as much control over our futures as lab mice.

The way it figured to work was this: if you flew 7, you could back up

Mission	Optimistic Sequence	Very Optimistic Sequence
Command Module only in earth orbit.	Apollo 8	Apollo 7
Command Module and Lunar Module in earth orbit.	Apollo 9	Apollo 8
Command Module and Lunar Module in earth orbit. Lunar mission simulation with a 4,000-mile apogee.	Apollo 10	Apollo 9
Command Module and Lunar Module in earth orbit. Lunar mission simulation.	Apollo 11	
Lunar landing attempt.	Apollo 12	Apollo 10

10. If you backed up 8, you could fly 11, and so on down the line. The system was subject to change and would, in fact, be modified on several occasions. Coming out of the meeting we knew, when it came time to announce a lunar landing crew, it would be desirable—if not essential—to have all three pilots flight-experienced and through the Apollo training cycle at least once before.

So there it was, down to the eighteen finalists, although several of us—Schweickart, Anders, Eisele, and myself—would be making that qualifying flight a little late in the game. Two of the crews—the Apollo 7 backup of Stafford, Young, and Cernan; and the Apollo 8 backup of Armstrong, Lovell, and Aldrin—met all the qualifications. One crew, Seven, had two rookies, Don Eisele and me. (What had Wally Schirra done to deserve that?) But the Apollo 7 mission was the only one cast in concrete and it would be the first out of the chute. Conspicuous by its absence in either plan was what came to be called the Apollo 8 mission (Command Module only in lunar orbit).

Looking over our group of eighteen, I didn't need a computer to realize I was the longest shot in the room. Low in experience, popularity, and political instincts, it had been a long shot just for me to be there. I simply shut off that vision of myself making the first lunar landing. The first view of the planet earth from an altitude above 10,000 miles definitely seemed a more attainable goal.

That one of course fell to the Apollo 8 crew.

Frank Borman is a man capable of success in whatever he attempts. He is a doer and a born leader, but he can be an imperious martinet to those who work for him. He seemed to glow. He wore the smug look of one who knows he will be the anointed one to do "it," whatever "it" may be. In many ways he was like Wally Schirra, but smarter and more effective in pursuing his objectives. He made decisions quickly and rationally.

Jim Lovell is an easygoing fellow who delivers uninspired but competent performances. I was first introduced to him as "Shakey," a nickname bestowed on him early in his flying career by his friend, Pete Conrad. Jim's career at NASA was marked by his singular ability to be in the right place at the right time. It carried him to several space endurance records and, in the process, the distinction of being the first man to fly on four different space missions.

But Jim didn't win every hand. The break that made his flight on Apollo 8 possible (Mike Collins' back problem) broke the rotation and cost him a seat on the big one. He would have flown to the moon on Apollo 11 with Neil Armstrong's crew. Instead, the Apollo 13 spacecraft which he commanded miscarried and his planned lunar landing had to be aborted.

The third member of the Eight crew, Bill Anders, was bright, dedicated, and one of the few people I ever met as stubborn and as serious as myself. A good Catholic with six children, he liked to tell his friends, "In my family there is a king, queen, and serfs." The two of us shared another distinction: we were the only two astronauts to carry the title Lunar Module Pilot with no Lunar Module to fly.

Apollo 8 was another precise, nearly perfect, giant step to the moon. Its unique accomplishment was in removing any psychological barriers associated with escaping the earth's gravity—if, in fact, they ever existed. Everyone who could squeeze into Mission Control and in the viewing room behind—some two hundred of us—watched tensely as Apollo 8 disappeared behind the moon for the rocket burn that would commit them to lunar orbit. For nearly an hour, while they passed behind the moon and out of radio contact, those on earth could only sweat it out. No human power could reach them; no ear could hear their call. And then, at the precise moment they were to appear on the eastern rim, the CapCom tried to reach them.

"Apollo Eight Houston . . . Apollo Eight Houston, over."

For seconds that seemed like minutes there was silence. Then Jim Lovell's voice confirmed they were in lunar orbit. "Go ahead Houston, Apollo Eight."

He then provided us with man's first impression of the lunar surface: "The moon is essentially gray—no color—looks like plaster of paris—sort of gray sand. . . ."

Anders added, "Looks like a sand pile my kids have been playing in for a long time—it's all beat up—no definition—just a lot of bumps and holes."

It was Christmas week 1968, the second greatest Christmas story ever told. Borman's crew pulled off the public relations coup of the century when they gave a shared reading of the first chapter of Genesis on Christmas Eve. Who could help but be moved when Frank began the first of ten verses from the Bible: "In the beginning God created the heaven and the earth . . ."

That moment established Apollo 8 indelibly in the minds of millions of TV viewers around the world as the Christmas mission. Today much of the public thinks of it as the first flight in the Apollo series, an impression that—I am frank to admit—causes me an occasional twinge. It was a beautiful touch and took a selling job by Borman to get Anders to read from the King James version of the Bible. Lovell's participation seemed a little incongruous, but he came through loud and clear.

Gossip at the time had it that Borman got the idea from a friend of his with the CIA. That seemed little more than an amusing footnote. Had it been six years later, of course, there would have been an investigation. The suggestion for the reading actually came from Si Bourgin, of the U.S. Information Agency, who correctly predicted the world-wide reception would be overwhelming.

The Scripture reading was universally well-received, with the exception of Madalyn Murray O'Hair, the militant atheist from Austin, Texas. She filed a lawsuit against NASA, or the government, or both, demanding equal time, or something.

On the tenth and final trip across the back of the moon came the most suspenseful moment of the space program to that time. Shortly before reestablishing contact it was necessary to burn the service module engine for two and a half minutes if they were ever to return home. Talk about a pregnant pause—that was it. Waiting for radio contact at the proper acquisition time, we all held our breath, and when they appeared on time with the spacecraft tracking for home, a loud and spontaneous cheer went up that would be repeated each time this scene was re-enacted in the years following. It seemed only minutes later that Borman reported "Lovell is already snoring."

In the control room we were all laughing then in that nervous way caused by excitement. Borman, Lovell, and Anders had hung their fannies over the cliff and had been able to bring them in on cue. It was not the cheap thrill of a circus stunt or of forcing oneself to dive off a thirty-foot tower or submitting to an impulse to walk the edge of a tall building, but rather the satisfaction of doing a dangerous thing professionally, under control all the way.

Apollo 8 also chalked up another first. After years of dedicated pessimism by the NASA doctors concerning the ability of man's body to withstand the physiological shock of zero g, they almost had their heyday at the beginning of the mission. For the first time a flight was close to termination because of on-board sickness. The problem had plagued the Russian cosmonauts, but we had never experienced any indication of zero-gravity motion sickness through the six Mercury and ten Gemini flights. Yet, shortly after leaving earth orbit on their way to the moon, Frank Borman became ill with nausea, vomiting, and diarrhea. The symptoms could have spelled real trouble.

Frank later attributed it to a reaction to a sleeping pill he had taken. Or maybe it was a twenty-four-hour virus. Mainly it was a mess. He wasn't throwing up on the floor because there is no floor in zero gravity. When you get sick in space it's three-dimensional. Lovell and Anders were worried about Frank, but they were also worried about living in a pretty funky spacecraft for seven days as well. Ground control didn't learn about Borman's problem until some ten hours after it struck, when the contents of the on-board tape recorder were played back on the ground. By that time Frank was coming out of it and no one wanted to terminate. They toughed it out.

NASA doctors had a running frustration trying to impress on us the potential for nausea in a spacecraft in which it was large enough to move around. We considered them prophets of doom and ignored them. (During the first day of Seven's mission, I had personally gone to some lengths to perform my own little experiment, going through all manner of head gyrations in an effort to approach the point of malaise.) None of us felt any motion sickness, even with Wally's cold, and we filed it away as only another boogie man raised by Chuck Berry's medical troops. It was to become a much more significant problem on Apollo 9.

Borman's crew was the first to receive the fullblown hero treatment since John Glenn, a world tour with international acclaim and honors. I was able to share the public's reaction two months later when Bill Anders and I were inducted into the Explorers Club in New York City. The Eight crew had carried an Explorers Club flag on their epic journey and Anders was to return it at this annual dinner.

We found ourselves on the program with an assortment of explorers who specialized in everything from studying the giant condors of South America to climbing Mount Everest to traveling by snowmobile to the North Pole. Bill and I felt like kids in the company of such seasoned campaigners, in the kind of room where you'd expect to find Vincent Price playing the organ while the draperies burned. It was Old World.

But I doubt that any of the members had ever been more deserving of honors than Bill Anders was that night. Bill presented the flag, then I listened with envy as he gave a travelog of his journey to the doorstep of the moon. The slide projector, clicking off photographs taken at various stages of the flight, created the illusion of an adventure shared. When he flashed a picture of the earth on the screen and said with studied casualness, "Here we are, one hundred sixty-five thousand miles out," I envied him as I had envied no man. It gave me pause, knowing that my friend Bill Anders could never be the same again. I'd have given anything to have been able to deliver the lecture Bill was making, but my assignment to Skylab had eliminated my chance of flying anything but an orbital mission in the future.

I felt then, and still do, that Apollo 8's role in our program to land a man on the moon was more form than substance—a work of art rather than a pushing back of boundaries. But it caused an emotional binge, a soaring of the spirit in a very real sense. In the context of man in space it was unique and wonderful, but if it had never occurred, Apollo 11 would not have been affected. Of course, that opinion is open to debate. Chris Kraft, the head of flight operations at the time, has stated it was the keystone of the Apollo program. And countless Americans, to this day, think of it as the first manned Apollo mission.

After Eight, there remained but one important piece of hardware yet to be tested, the most important piece: the Lunar Module (LM, pronounced lem). The first one off the line was christened Spider, an apt description, and took to the air with the Command Module (Gum Drop) on an unheralded mission early in 1969. It was to be flown by one of the better-trained crews ever launched. Jim McDivitt, Dave Scott, and Rusty Schweickart had been in training as a crew since January 1966 when they were assigned as the backup unit to Gus Grissom's ill-fated Apollo 1. Jim and Rusty were the first two assigned to follow the production LM's out of the factory, and a great chunk of their careers hinged on that early decision.

This crew would give the Lunar Module a thorough shakedown. To accomplish this they would be in the unenviable position of flying a spacecraft in earth orbit without a heat shield. The LM was designed for landing on the moon where there is no atmosphere and, consequently, no need for a heat shield. After separating from the Command Module in the LM, the only possible way back to earth was to rendezvous with their Command Module, which had the heat shield protection to withstand reentry.

The members of the Apollo 9 crew were looked upon around the office as fair-haired boys, and sometimes referred to with some cynicism as "the rah-rah boys." Through all the ups and downs and changing of gears early in the Apollo program, somebody up there always seemed to look out for McDivitt's crew. With personnel transfusions into Apollo from late Gemini flights, the pad fire, and mission changes due to hardware problems, some astro fortunes rose and others fell but one thing remained constant. From the time we started training for Apollo in early 1966: Mac's crew was assigned the first Lunar Module and, by hang, they would fly the first Lunar Module.

No one was more deliberate or meticulous than Jim McDivitt; no detail was too small if it affected his mission. He eliminated problems by working them to death. When he presented his case he had the facts—the antithesis of Schirra's patented emotional appeal. Most of Jim's contemporaries considered him the "anointed one" of the nine Gemini astros. Looking back it does appear he was given every chance, break, prerogative, and option, beginning with command of the second manned Gemini mission.

Rusty Schweickart would be the LM pilot and perform the first EVA of the Apollo program. He was bright, an idealist, at the intellectual end of our spectrum. As the only civilians in our group of fourteen, he and I spent a brief period after our arrival at NASA as pseudo scientists, sops to the National Science Foundation, which was then lobbying to get one of its own into space.

The third member of this trio, and one with nearly as much potential as McDivitt, was Dave Scott. He was the golden boy of our group, well-educated, athletic, with boyish good looks (and frequently an attitude to match). Above all, he was a fine aviator, and his career, too, seemed to have been made in heaven.

There was more reason to be apprehensive about the safety of the first Lunar Module crew than for the passengers on Apollo 8. If anything went awry, Jim and Rusty could find themselves adrift in space in a machine with no capability of returning to earth safely. In that unlikely event, Dave, flying the Command Module, would be called upon to make the first rescue and singlehanded space rendezvous. You can believe Dave spent one hell of a lot of time practicing in the simulator for such a contingency.

The Nine mission filled several squares, although Rusty's EVA turned out to be only a semblance of the original plan. This was due to the one major problem experienced during the flight. What had been debatable with Frank Borman on Eight was an indisputable fact with Rusty on Nine: he was sic. as a dog for four days. Only determination and drive enabled him to stick it out through some of the most miserable hours known to man. It was the granddaddy of airsick cases, at least until the Skylab launches in 1973.

Only those who have experienced it can fully appreciate what Rusty suffered, 200 miles out in space, weightless—or what it took to finish the mission as planned. They didn't have weeks or months to adjust, as Skylab would have. To the contrary, their first five or six days were typical of all the Gemini and Apollo missions; overloaded on the front end in case of an early termination. Not yet fully recovered, and still wary of further sickness, Rusty piled himself into his pressure suit, climbed out of the Lunar Module, and performed what he could of his scheduled space walk.

Nine was a crucial step to the moon, a well-executed engineering test flight. But rare is the layman who recalls there were two earth-orbital missions in the Apollo program.

Nine left its imprint on Rusty Schweickart. The motion-sickness experts in Pensacola, Florida, found him a cooperative guinea pig in their search for a cure. He threw himself earnestly into helping them develop a conditioning program for those with similar tendencies (although there is still no way to predict who might have them). When doctors eventually prescribed a head-motion exercise to be followed during the first hours in zero g, some blamed it for subsequent stomach "awareness."

Rusty's efforts earned him countless rides on a rotating chair at high rates of speed while making a terrific number of head motions. But I don't believe he was ever again seriously considered a candidate for future space flights. His role as a backup commander in Skylab held out the prospect, but it never came to the test. Later events—meaning a wave of zero-g sickness among the Skylab crews—indicated it would not have been a smart gamble. After his return from Apollo 9, Rusty never appeared all that enthusiastic to revisit the scene of his misery. The code required an effort and he made one. If it took a bit of his spark out, no one held it against him.

It was becoming apparent by now that the zero-gravity brand of motion sickness was very subjective. It remains a major physiological problem that must be solved if the dream of universal space travel is ever to be realized. Traveling to other planets without a solution will surely make the barf-bag manufacturers rich.

Even before Nine lifted off, the flight crew assignments through Apollo 12 were obvious. It hadn't been determined yet which mission would attempt the first lunar landing, but most of us were guessing that Twelve would catch the brass ring. It still seemed improbable that we could put together the success record necessary to let Apollo 11 go for it.

An Apollo family portrait would picture Nine as an engineering student standing between Eight and Ten, two strapping glamor-boys who traveled to the moon. Politicians advise in any group picture to avoid standing on the end because it is easy to be cut off and forgotten when the results are printed. That is where Seven stood. Apollo 10, next up, was a full-dress rehearsal of the lunar landing.

Just seven months after the flight of the Phoenix, our backup crew of Stafford, Young, and Cernan was ready to fly their own Apollo spacecraft. Apollo 10 carried a veteran crew. Stafford and Young had each made two Gemini flights and Cernan one.

Being a backup crew member is not the easiest assignment. The responsibility is a great deal more than just training man-for-man and as a unit to fly the mission. It is a support role to the prime crew, handling tasks that the prime crew cannot (or chooses not to). Human nature being what it is, there is a tendency to follow their own downstream mission plans and hardware as closely as possible. They must also train for their role in mission control during the flight. On the one hand, getting your prime crew launched demands everyone's concentrated effort. On the other hand, the key to your future is more dependent on how well your own anticipated mission goes. It was small consolation to get your prime crew launched only to find one's own mission had gotten completely screwed up before you arrived on the scene. Conscientious efforts at backup training were more than justified if one had to step in and fly the mission, but for this to happen one must wait for his buddy to break a leg, catch the measles, or worse. You couldn't help but feel like a vulture occasionally.

Protocol and historical precedence tend to restrict a backup crewman from knocking heads with the prime crew over an issue, in spite of the fact that the backup crew may be at a higher level of training or experience. When these three backed us up on Seven, their thoughts were frequently on flying Ten to the moon and back. There were many times when I wished that Stafford, as backup commander, had pursued his views more energetically when they differed from Wally's. He had lived through two tours of the "even strain" with Wally on Gemini and served as Wally's backup for two years on Seven. In his backup capacity, he seldom chose to ruffle Wally's feathers, even when he agreed with Donn and me and knew we could use all the help we could get. But Stafford is a politician and it was his counsel that eased the pent-up frustrations of Donn and me during the months of working with Wally.

Young, who backed up Donn, was old reliable. There was nothing spectacular about his work, but John would do anything he was called upon for, anywhere in the country, on whatever schedule was necessary. He was one of the most uncomplaining, hard workers I have ever known; thorough in his analysis and no-nonsense in his conclusions.

When asked in mid-1967 to evaluate a beefing up of the spacecraft couches to survive a land "landing" (the couches not the crew), he concluded, "Make no mistake, we will be lucky to survive a land landing in this beast. A NORMAL LAND 'LANDING' CAPABILITY DOES *NOT* EXIST IN THIS VEHICLE. To provide it will cost us two years and many megabucks. It is recommended that we admit this fact and that crews push others to quit adding weight internal to the vehicle that cannot begin to improve crew survivability in a land 'landing.' And let's go *FLY*."

Gene Cernan was the other end of the spectrum from John. He met the schedule and filled the squares, but we never got the impression he stayed up late worrying, or working on the problems. The job was so stimulating and satisfying that most of us were rarely able to put it completely out of mind. Gene never seemed to have this problem. He is the most gregarious, outgoing, sociable astro of the bunch, and is in his element as the partynaut in Hollywood or Washington, D.C. Gene and John kept their counsel with Stafford more than with Wally, and operated more in the role of the Apollo 10 prime crew than the Apollo 7 backup.

With one wild exception, Ten was a perfect test flight. For the first time in three trips no one got sick. Every phase of the LM mission went without fault, except for the final ten minutes of powered descent.

Their Lunar Module trajectory took Stafford and Cernan within 50,000 feet of the surface of the moon where they were to simulate an abort with the ascent-stage engine. At separation from the descent stage, a single switch left in the wrong position caused the spacecraft to take off on a wild, unscheduled, head-over-heels maneuver. In an operation where everything has been planned to a gnat's eyelash, the unexpected can really scare the pee out of a fellow. With all the conversation going live to the ground, Cernan let out an expletive that was not deleted. In fact, it was heard around the world. "Sonofabitch," Gene blurted, "what the hell happened?" Once more the unexpected and unimportant became the trademark of the mission. Apollo 10 was to be remembered as the flight where an astronaut swore while the whole world listened. I never had the heart to ask who missed the switch.

Back on the ground, Apollo 10 was serving as the swan song for the youngest of the original seven astronauts, Gordon Cooper. Frustrated, fed up with getting what he took to be the short end of the stick, and weary with NASA's continued intrusion on his private life, Gordo too often seemed to be just going through the motions. Many of us had viewed his case with sympathy, but at this point he lost the support even of those in whom the loyalty instinct died hard. He was getting a raw deal, but that does not excuse giving less than your best.

With Cooper on the Ten backup crew were Donn Eisele, who was always quick to adopt the behavior pattern of his leader. Therefore, he was also operating at less than full speed. Ed Mitchell, the rookie of the crew, was the newest, the youngest, the brightest, and the hungriest, and he went at the training grind as though it were a contest. His performance did not go unnoticed and secured him a seat on Al Shepard's Apollo 14 crew.

With the launch of Ten it became official that Apollo 11 would attempt the first manned landing on the moon. The entire country was examining the crew of Armstrong, Aldrin, and Collins as the products of an exhausting search among the astronauts to find the crème de la crème—to settle on those individuals among all the rest who would be most apt to perform the best job on the most demanding mission in history. They had good credentials and it would be nice to reassure everyone that there was a science to selecting the crew for such a momentous event. But it just wasn't so! In the end it was simply the rotation, not competition, that decided which mortals would be the first to land on the moon.

They were not picked by computers, nor by accident, but by a human process that fell somewhere in between. Deke Slayton always insisted that any of his boys could do the job, that the crucial selection had not been for crew assignments, but to qualify for the job in the first place. To Deke we wore the same label: ONE EACH, MARK IV, MOD 3, ASTRONAUT. I think Deke believed it and I believe Deke. There were rare exceptions, but in those days it was essentially correct.

In theory, everyone started even. Fate, politics, and performance, more or less in that order, took over from there. It's politics when the three active Mercury astronauts could take their pick of the Gemini missions, leaving the generally better-qualified Gemini astros to work in behind them. Grissom the first one, Wally the glamorous first rendezvous mission, and Cooper gets the best of the rest of the early missions. In the jockeying for position at the start up of the Apollo program politics took a back seat to fate.

When the *original* Apollo flight schedule (in 1966, before the fire) got underway, half of those who would eventually fly on the first six missions were still tied up on tail-end Gemini crews. Training for the two programs had overlapped by more than a year. But when the Apollo 1 fire caused a twenty-one-month delay, all of the Gemini crewmen, including Armstrong and Aldrin, became available for the restart of Apollo.

Even the fire had less effect on the order of things than an earlier shake-up at the end of the Gemini program. When Elliott See and Charlie Bassett, the prime crew for Gemini 9, were killed in St. Louis, the impact on flight crew assignments was the most significant ever experienced. With the first two crew assignments in Apollo already made, astros who had been counted upon to follow in line suddenly became unavailable because of Gemini crew shake-ups, while others who had been expected to be tied up on Gemini for another six months came on the market.

Stafford and Cernan, who were scheduled to fly Gemini 12, the last

Gemini mission, moved up to Gemini 9, and Stafford was able to make an early bailout to join Apollo. Jim Lovell and Buzz Aldrin, stuck in dead-end spots backing up Gemini 10 (with no Gemini flight to go to from there), moved to Gemini 9 backup and then inherited the last Gemini mission. This twist of fortune gave Buzz the necessary flight qualifications to be considered for the first lunar landing. The subsequent availability of many early Apollo crewmen could be traced to this accident.

Ours wasn't the first business where members of a small group found their careers enhanced by the misfortune of a respected peer. Though we never got used to it, we never looked back, either. If we had, we would have noted that while Armstrong was assigned a dead-ended position on Gemini 11 backup, his Gemini 8 crewmate, Dave Scott, had been moved directly into Apollo on Grissom's backup crew. If Armstrong already had a ticket to the moon, it would have made better sense to give him the Apollo experience. So which of the two was the fair-haired boy?

Hardware problems caused still other changes and delays. From the beginning, the first Lunar Module had been the responsibility of Mc-Divitt's crew. Borman's crew had the second, and they were assigned to the third Apollo mission, which would have taken them to the highest altitude above the earth ever attained: 4,000 miles.

No one was very surprised before Apollo 7 when it became clear that the launch dates for both the McDivitt and the Borman crews were threatening to slip due to delays in manufacture of the first Lunar Modules. One solution to maintain the regular launch schedule was to fly a second Command-Module-only mission, and slip each of the Lunar Modules back one mission. When management finally bit the bullet on that decision, McDivitt was offered the option of still flying that second mission or sticking with the first Lunar Module and dropping back to the third mission. He opted to stick with the LM they had lived with for two years. It was a characteristic response by the single-minded McDivitt. *Afterwards*, he returned to the office to ask Dave and Rusty for their opinions.

Borman, more of an opportunist than Jim, went enthusiastically after the rumored possibility of a circumlunar flight. So, months before a decision was made to attempt a landing with Apollo 11, we had a crew line-up that looked like this:

Apollo 8. Prime: Borman, Collins, and Anders
Backup: Armstrong, Lovell, and Aldrin
Apollo 9: Prime: McDivitt, Scott, and Schweickart
Backup: Conrad, Gordon, and Bean

Now we can play the what-if game. *If* the LM hardware hadn't slipped, or *if* Jim McDivitt had exercised his option to fly the second mission, then the following events would have taken place: Borman's crew, with Mike Collins recovered from a back operation, would have flown an earth-orbital mission with Apollo 9; their backup crew of Armstrong, Lovell, and Aldrin would have rotated to Apollo 12, the second lunar landing; McDivitt's crew would have gone around the moon on Apollo 8; and Pete Conrad and Alan Bean, from their backup crew, would have been the first two men to set foot on the moon. The what-if's for the See-Bassett and Apollo 1 accidents are too much to attempt.

In early 1968, when all this was taking place, it would have been interesting to know the odds on Apollo 11 winding up as the first moonship. That would have been easy compared to parlaying the See-Bassett accident, the Apollo 1 fire, hardware schedule slips, and several individual physical problems to predict which astros would come up with the Apollo 11 brass ring: Jimmy the Greek, where were you when we needed you?

By the time the mission got off the ground, Neil Armstrong, Buzz Aldrin, and Mike Collins had lived for six months with the historical significance of what they were about to attempt. They were an interesting study. Mike had undoubtedly adjusted smoothly to his role and contribution to the mission. Neil could paste an Alfred E. Neuman smile on his face, content in his assignment, and worry about what he should say on that first small step, while Buzz must have been seething and bubbling inside like a caldron.

Professionally, they made a good team. Mike was pragmatic, Buzz was the theoretician, and Neil was the one with a feel for the hardware and the cool judgment. Everything Neil did was low-key. He flew the X-15 rocket plane early in its development, yet I never heard him mention it. He never seemed to get involved emotionally in an issue, and when he spoke, his tone and delivery gave the impression that his words had been rehearsed. His face has a softness, a pinkness, that will keep him looking younger than he is. He was one of the least athletic of the astronauts; if he felt the urge to exercise he'd lie down until it went away. To those who knew him, Neil wasn't the type to become involved in the superspectacular. Whatever he did, he seldom made mistakes. He was the right man for mission commander.

Like Neil, Mike Collins rarely got involved in office politics or in the issues the rest of us campaigned for. He had a casualness that struck some as laziness, but he was a better technician than he claimed in his book. His EVA on Gemini 10 was the most effective of the program. Mike's one emotional involvement came as a leader of the thriving antidoctor clique among the astronauts, resisting any attempt to use us as guinea pigs. He was also the best handball player in the office, a ranking I frequently challenged . . . unsuccessfully.

Buzz Aldrin, the LM pilot, was a puzzle to most of us from the time our group first came together. His dedication to the science of space rendezvous was almost legendary. He was personally responsible for many of the initial steps and later innovations that brought rendezvous to the level of success it enjoyed prior to Apollo 11. Many of the essential aspects of the mission were dependent upon computer software, another of Buzz's fortes.

It was obvious that Buzz was doing a slow burn when the rumors began that Armstrong would take the first step on the moon. He seemed to have presumed that he would be the first man out, possibly because an early version of the checklist called it that way. There was never any mystery to the process by which Neil was accorded that honor. The sinister-influence theory—the idea that forces in Washington wanted a civilian to do it—struck me as ludicrous. Neil was the command pilot, a member of the second generation of astronauts. Seniority, the pecking order, and all the sacred cows required that he go first—if going first was the desirable thing to do.

Aldrin's efforts to checkmate Armstrong, which took the form of behind-the-scenes lobbying, much of it aimed at Tom Stafford, irritated many of us. Stafford was deeply involved in the mission planning and Aldrin, as the Lunar Module pilot, was pushing for the final mission plan to show the LM pilot stepping out of the hatch first. (It didn't.)

Meanwhile, Aldrin's dad was trying to shake a few trees in Washington to correct the "injustice" of Buzz not stepping out first, no doubt to the embarrassment of Buzz. It wasn't the first time that the elder Aldrin's good intentions had muddied the waters of his son's career. Before Gemini 12, the old colonel had injected himself into the question of the spot promotion which then *Major* Aldrin would receive after his first space flight. It has become a tradition after a first flight for the parent service to advance their military astronauts to the next higher rank. Those efforts won no friends for Buzz, nor did Buzz's own campaign to beat Armstrong to the moondust. The whole issue resulted in at least one confrontation between them, but it did not visibly affect their ability to carry out a beautiful mission. His father was still writing letters to the editor months after the flight while Buzz developed a stock response, reminding people, "We both landed at the same time."

The fact is, Neil's footprints are ten years older now. That may be a speck of sand in the hourglass of time, as the saying goes, but it is fortunate for the country. The reason is basic. No one knows how Buzz would have handled that pressure, including Buzz himself. But Neil did so with dignity and grace and then retreated to the quiet, scholarly life he had always wanted, as a faculty member at the University of Cincinnati.

The scene at the Cape on the morning of July 16, 1969 resembled a carnival more than the beginning of man's greatest adventure. What a vast difference it must have been from the sailings of the fifteenth- and sixteenth-century explorers, or the dignified departure of Lindbergh's *Spirit of St. Louis* from fog-shrouded Roosevelt Field, Long Island, forty years before. For this trip most of the world was watching. In fact, most of the world seemed to have descended upon Cape Kennedy, squeezing into the firing room and overflowing the VIP bleachers. In living rooms all over America, parents were telling their kids, "Shut up now and watch. The men in that rocket ship are going to the moon," the words sounding not quite believable even to their own ears.

But we believed. Regardless of what else they had going at the moment, most of the astronauts found some excuse to be at the Cape with a million others to watch the United States send its missionaries to another world. At lift-off we pushed with our hearts and our heads, as we had for all the other missions, but we knew that nothing new and earthshaking would take place for three and a half days. Until the actual landing, every event, every maneuver, had been performed previously from one to four times.

We reserved our apprehension for the landing, the only thing that hadn't yet been demonstrated. That's what created our suspense. For the first time, Apollo would go the last 50,000 feet. At Mission Control, shortly after 4:00 P.M. on Sunday, July 20, 1969, we waited as that final descent began. We watched displays that even the crew could not see and listened silently for sounds that had yet to be uttered. If a man so much as cleared his throat a hundred other voices shushed him.

We thought, "They are actually going to pull it off." And a companion thought flashed through my mind: "Men that I know are landing on the moon." I looked around the room and realized that each of us was individually immersed in the moment and what it meant. It was what we had each been working for all those years.

Sitting there in our Walter Mitty world of Mission Control, the private thoughts of each man were penetrated by two events in the final descent. As their space bug, Eagle, plunged rapidly toward the surface of the moon, Neil and Buzz began getting a persistent alarm from one of the computers. They were being read on the ground also, and each time Aldrin sang out, "Another twelve-oh-one alarm," Cap-

Com would come back with, "Roger, alarm." It was the signal for a computer overload—which could play hell with its job of calculating the LM's altitude and rate of descent and it was conflicting with radar data. Simply put, to land safely they had to know how fast they were descending.

The alarms were creating doubt, one of a pilot's deadliest enemies. Confidence had to be restored that the problem would correct itself or that the landing could be accomplished in spite of it. There was no time for discussion. A young engineer named Steve Bales concluded that the computer would clear itself in a few moments. If he felt any hesitation, his manner didn't betray it. In the stillness of the room he nodded and said, "Go, flight," and 6,000 feet from the moon the Eagle was cleared to continue. It is no exaggeration to say that, at that moment, the landing could have been aborted, unnecessarily. The quiet "Go" of young Bales was undoubtedly one of the most crucial decisions ever made in Mission Control.

It is relatively simple to make a life-and-death decision when it is your own life in the balance; making that decision when someone else's life is at stake is a different thing entirely. Bales would confess later, "I was scared to death." He had the knowledge and the guts to make his decision, and for that judgment and decisiveness he would be awarded a Medal of Freedom at a White House ceremony standing alongside the crew of Apollo 11.

With that kind of tension, and the packed viewing room of VIP's looking over their shoulders, the flight controllers were their usual picture of efficiency and decorum. The viewing room is not unlike the observation level in a hospital when a famous surgeon is operating. People walk softly in and out. The room is filled with professional respect.

Even during the deadest midwatch there is rarely any trivia heard over the two communication loops monitored in the viewing room. Yet within two minutes of landing there was brief consternation and then laughter at a wisecrack heard over the internal loop (within the control center only). Unable to resist the purely American instinct for a gag at the most solemn and critical moments, a young engineer had injected a comment that was cynically in context, yet, in the pressure of the undertaking at hand, outrageously funny. "The president is go for landing," came loud and clear over the speaker. It was triggered either by Mr. Nixon's eagerness to get into the act (he was standing by, via telephone, to talk to them on the moon) or the fact that everyone (including the president) was but a spectator to what was transpiring in *our* world at that moment. While the world watched, Neil and Buzz made their historic landing. Neil's transmission sent goose bumps around the globe: "Houston, Tranquility Base here. The Eagle has landed."

Charlie Duke, the CapCom, replied, "Roger, Tranquility, we copy you on the ground. You got a bunch of guys about to turn blue. We're breathing again. Thanks a lot."

Make no mistake, the significant event was the landing, not man's footprint in the lunar dust which came several hours later. That was show biz. The first footprints were symbolic, but when that ugly, spidery contraption—"like a praying mantis," as Mike Collins described it—touched its claws to the moon, man had succeeded in one of his most difficult undertakings. It came at the end of a long and tenuous chain of man-made circumstances and occurred nearly two years late by the original timetable. Special credit is due the amazing pieces of machinery that got us there and the thousands of brilliant minds that conceived them. Many call it luck. I have said we were "lucky" to do it in only five flights, but there was no luck involved in getting there. It was the best-documented example in history of science and skill overcoming superstition and ignorance.

As Neil and Buzz were shaping that chain of circumstances to a touchdown with barely twenty seconds of fuel remaining, they were experiencing the very essence of what living is all about. While science and skill gave them the tools to control their own fate, they, nevertheless, were walking the tightrope that separates life and death. Many men walk that tightrope during their lives, but this time they were coming down the chute to an adventure so fine, it was out where the meter doesn't even register. With the slightest misstep they could find themselves in oblivion as martyred heroes.

All of us who watched and waited thrilled at the vicarious experience of those first two humans walking on the moon. When it came time for lift-off, Mission Control was packed once more, but the atmosphere was not the keen, almost metallic sense of anticipation that grew as the landing neared. Now came the collective urgency of getting them back. It was as though "we" had put them into their hole and now, by God, "we" had to get them out by our concern and prayers. We strained as the Eagle ascended from the moon. And they did climb out because everything worked as advertised, including the crew. While Neil and Buzz were still in the boost phase of their departure we were faced with the fact, incredible as it seemed, that the United States had actually pulled it off. We put a man on the moon within the time promised and nearly within the cost advertised. When it came right down to it, that fifth giant step was easier than we had expected. Maybe, more correctly, the crew and the hardware made the impossible look easy.

In eight short minutes the LM was in lunar orbit and Mission Control began clearing out rapidly. What the hell, the rest had all been done before. We could watch it all on television and have a drink while doing it. Frankly, we assumed the remainder of the mission would be a piece of cake for the guys—relatively speaking, of course. How quickly we take for granted what so recently seemed impossible.

13

A Beaten Path

THE NATION suffered a letdown after the first moon landing, not unlike what a baseball team and its fans must feel when the pennant has been clinched but there are still games to be played. The old team spirit may get bruised but it's a good time for personal goals and controversy and touches of temper.

That was the sense of what happened to us after Apollo 11. This first landing, the beginning, had become an end in itself. What were we to do as an encore? Space agency planning had produced nothing more than a series of ten, basically me-too missions to the lunar surface over a period of some approximately thirty months. Each would go to a different site, always a compromise between scientific curiosity and mission risk. Mobility on the moon's surface would be increased and some variation in the science package would be introduced with each landing. To the American people the moon was likened to the old story of the elephant investigated by blind men: whatever they grabbed, wherever they touched, they got an entirely new impression. The moon was being sold as our great gray elephant.

Once Armstrong's crew brought home their color photos and rocks and moondust, we really didn't figure to get all that many surprises, and Congress, it turned out, wasn't all that interested in unlocking the secrets of the solar system. The space program wasn't selling, which was why the public relations boys began to place more emphasis on the benefits to everyday living that would result from our space discoveries. It was like trying to convince racing fans that the purpose of the Indianapolis 500 is to develop safer passenger cars.

Congress, too, must have concluded that the missions were repetitious, and pressure from that quarter forced NASA, in late 1969, to cancel Apollo missions 18, 19, and 20. There were a lot of wounded squawks when the cuts came and two of the loudest were traced to the astronauts and the geologists. Up to then the geologists had pretty much treated the moon as their own private sandbox. The astros, of course, saw nine seats being deleted from planned space voyages when their life's work was to fly in space.

In my opinion, the cuts were a sensible, prudent move. This country's first orbital laboratory suffered for five years, while the lunar landings were draining off the lion's share of the space dollars. Had the lunar missions been reduced to, say, five, Skylab could have been in the air two years earlier.

There was never any question about flying Apollo 12, however, and it was born in a thunderstorm in November 1969. While Seven had splashed down in the worst weather ceiling ever (200 feet), the record for bad weather at lift-off belongs to Twelve. Visibility was barely two miles, in a hard rain, and by the time the rocket had reached 500 feet only the fire from the engines was visible, burning its way through the clouds. At 1,000 feet the rocket was struck by lightning and the spacecraft was automatically switched to backup battery power. Thirty seconds later it was struck again, and the on-board computer began displaying garbage from the navigation platform. It was not exactly a happy-go-lucky lift-off for Pete Conrad, Dick Gordon, and Al Bean. One more bolt and the boys could have been singing "Nearer My God to Thee."

Pete Conrad is a compact five foot six, but he had that special presence that stood out in a crowd. Early in the program we thought he was a half-million-dollar-better bet to be the first man on the moon than anyone else in the office. That was when we were working with the Grumman engineers on a "super weight improvement program" for the Lunar Module. NASA was willing to spend $30,000 a pound on any change that would lighten the Lunar Module, and Pete weighed a good fifteen pounds less than any of the rest of us.

He was a Princeton graduate and a career naval officer. As an instructor at the Naval Test Pilot School, he had counted among his students his two crewmates, Bean and Gordon. Pete was bright without being an intellectual, in good physical condition without being a jock, and an independent and outspoken thinker without ruffling feathers (a trait I envied). He even raced sport cars, without catching the flak that Gordon Cooper did for the same hobby.

Pete had the confidence and judgment to be an ideal spacecraft commander. The closest thing around to Pete was Air Force Captain Charles C. Charles, the "Terry and the Pirates" comic strip character. If you were going to make a movie about an astronaut, Pete is the guy you would pick to play the lead.

At least one observer was willing to bet that Conrad would be the first man on the moon. Gerry Morton, of TRW, wagered $100 on it with a friend, and later told Chris Kraft about it. Months later, after the first of the Apollo flights, Chris was at Gerry's house for dinner. Casually, he said, "Looks like you won your bet, Gerry. Pete will probably be on the first lunar landing."

The timetable then pointed to Apollo 12 as the earliest possible attempt. But as the early missions continued to perform without a serious hitch, Kraft, George Low, and Wernher von Braun became convinced that the landing could and should be attempted on Apollo 11. They believed it enough to carry the plan to Dr. Tom Paine, the administrator of NASA. Paine bought it and, as developments were to prove, their judgment was correct. Of course, in the process, Pete lost out as the first man on the moon.

Dick Gordon is short, dark, good looking, with a raw sex appeal that overshadowed most of the rest of us. He is gregarious and cocky, in sharp contrast to his crewmate Bean.

Alan Bean was a thoughtful, cautious type, with the gaunt, balding looks to match. He had a body hardened by college gymnastics and was still in good shape. He was famous around the office for stubbornness in pursuing his own ideas, generally in the area of crew comfort and procedures. He was a native Texan, from Fort Worth and the University of Texas. His instincts often clashed with the military system, and while he could always be expected to drop the public pursuit of an idea at a senior's suggestion, he would continue to push for it in private.

Pete once said, "When you think you have the door closed on one of Bean's ideas, you had better keep running around to all the side doors to see that he's not slipping it under one of them."

It was a good crew.

While they were fighting to recover from the lightning strikes and save their mission, Sue Bean and some of the rest of us went to a birthday party for Vice President Spiro T. Agnew, the arrangements for which were made by Al Bishop, the RCA representative at the Cape who would soon go to work for Robert Maheu and the Howard Hughes organization. Al would remain a regular at each launch, always escorting a mob of people attending their first space shot. He also worked on occasion with the Secret Service, as an advance man for the vice president's visits around the country.

The party was to be one of those small, intimate affairs for Agnew's retinue and the astros, but it grew to resemble the cast of *Ben-Hur*. To find a place roomy enough to handle the crowd, and also assure some privacy and security, Bob Maheu stepped in to help. He shelled out $8,000 to charter a 120-foot yacht that was sailed from Miami for the party and docked at Port Canaveral. Once on board Agnew talked politics and Vietnam and played the piano while the rest of us lounged about and sang. He and his wife Judy were relaxed and delightful company. Years later I thought of that party, sadly, when Ted Agnew resigned from office, a ruined figure.

At Mission Control that night they were wrestling with the question of whether Apollo 12 was a wounded bird. For the first time in my recollection, both flight and mission directors began to shed their ultra-safe, conservative approach. They elected to allow Twelve to continue, while assessing the impact of the lightning strikes. Given a string of unqualified successes, they were moving closer to the crew's charge-ahead attitude. Twelve went on to one of the most impressive technical achievements of the program.

Several years before, the NASA unmanned program had landed a Surveyor spacecraft on the moon for the purpose of sampling the soil and making on-the-spot analysis for transmission back to earth. That now-dead spacecraft was resting in a small crater nearly a quarter of a million miles away. One of Twelve's objectives was to return to that spot to determine the effect of time and exposure on the Surveyor and to confirm the conclusions about the area reached on that earlier probe.

It was a real challenge to determine the precise location of the Surveyor and then to land within walking distance of that spot, all from a distance equal to ten times the circumference of the earth. It would be comparable to dropping a grape into a Coke bottle from the top of the Empire State Building.

The general area where the Surveyor had landed was known, but there were hundreds of small craters in the region, any one of which could be sheltering it. Before its demise, the Surveyor had transmitted 360° panoramic pictures of its surroundings. What followed was the kind of celestial detective work that made the space effort the success it was.

A geologist at the U.S. Geological Survey in Flagstaff, Arizona, studied these and other orbital photographs of the area and he selected those craters which could conceivably have been the nesting place of Surveyor. He then, painstakingly, reconstructed what might be seen, for 360° around, from each of these sites. It was then a matter of matching the photographs from Surveyor with the artist's conception of the view from the crater. The landing site was a bull's-eye, with Pete bringing the Lunar Module down within sight of the Surveyor.

Conrad's words were less than historic but in character when he became the third human being to set foot on the moon. Only Neil Armstrong knows how much thought and rehearsal went into his memorable quote, but I dare say Pete Conrad gave little thought to his irreverent paraphrase of it. Having safely dropped from the last rung of the ladder, which was still several feet short of the surface, he gulped and said. "That may have been a small step for Neil, but it was a giant leap for Pete."

Any member of the Russian space, military, or political hierarchy should have viewed that accomplishment—pinpointing an object and flying to it from a quarter of a million miles away—as second only to a successful lunar landing itself. The military implications of such a guidance capability are obvious.

A new wrinkle was introduced for the Apollo 11 postflight routine— quarantine. The crew of Apollo 12 was the second to undergo this experience in the Lunar Receiving Lab, an exercise that was a waste of both time and tax money. It was a medical boondoggle by the U.S. Public Health Service, an attempt to get a piece of the lunar action. To justify it, the medics had to first raise the fear that some germ, or some alien particle, would attach itself to the moon walkers and, after hitch-hiking to earth, contaminate our planet. It was "Twilight Zone" stuff.

The twenty-one-day quarantine of the Apollo 11 crew had been notable in the fact that by the time they checked into the Lunar Lab, the story of their adventure had already preceded them. When Mike Collins settled into his bunk, sitting at bedside was a paperback copy of *We Reach the Moon: The Story of Apollo 11*, by John Noble Wilford.

It was easier to send things in, of course, than to smuggle them out. There were several "accidents" in the first use of the lab's biological barrier, and a parade of people were exposed to the quarantine area. Each "victim" was dutifully confined on the other side of the barrier to become a part of the growing in-group. The list included one young lady technician whose motives we jokingly questioned. By the time the three-week sentence was served, the place was just about filled to capacity.

The crews took the quarantine with relatively good humor. "There wasn't a whole lot to complain about," summed up Alan Bean, "and if there was we couldn't do anything about it, so you might as well relax and enjoy it. Besides, it wasn't a bad way to get our postflight obligations out of the way. Our debriefings and pilot's reports got finished much more efficiently without the home and office routine."

Eventually even the diehards admitted that exposure to the moon posed no threat to humanity. No germs. No plagues. No blobs percolating in some scientist's back closet. After Apollo 14 the crew quarantine portion of the Lunar Receiving Lab was quietly converted to more useful purposes.

Meanwhile, back at the office, speculation continued around Apollos 13 and 14. Some serious maneuvering was in progress which resulted in Gordon Cooper bowing out to a desk job and Al Shepard becoming America's first *and* (briefly) last man in space; Jim Lovell would become the first to fly four missions.

In the usual order of things, Cooper's crew, having backed up Apollo 10, would fly Thirteen. Lovell's unit, backing up Apollo 11, would fly Fourteen. Keep in mind that even while training for a mission, the prime crew feels a certain anxiety until after the public announcement of the crew's identity, four to six months before the flight. When an unusual delay held off the Apollo 13 announcement, most of the astros began to suspect that something was afoot. The gamblers in the office were all betting that Shepard would be taking the ride.

No one argued with Al's right to fly again, but the objection to the way he was going about it was nearly unanimous. The details began to surface when the plan started to fall apart at the seams. Shepard's rare ear problem, of course, had restricted him from space flight since 1963. He was allowed to fly NASA aircraft only with a second qualified pilot on board, and for a man of Al's temperament, ambition, and ability, this kind of dependence had to be a helluva blow to his pride. Since 1964, his duties as chief of the Astronaut Office didn't take up all his time and energies; the rest had been channeled into a few personal goals: banks, real estate, and other business activities around Houston.

Since his fifteen-minute suborbital flight, Shepard had carried the title of "America's first man in space." Perhaps *tolerated* is a more apt word, in the same way Deke Slayton tolerated such references as "one of the Original Seven" and "the man who trained the men who went to the moon." While Al watched crew after crew taking the big ride, he must have felt like the guy left behind to guard the sheep while the hunters went after the fox.

In 1969 he saw the opportunity to remove any question of his ability, real or imagined, in his or anyone else's mind, and he grabbed it. In Los Angeles, in a new and risky operation, he had his ear problem corrected. The surgery was kept secret, but we soon noticed that Shepard was flying the T-38 without a second pilot. He stepped up his physical fitness program and began to disengage himself from his outside businesses.

Some of us got the message that Al was back and he was going to the moon. The next opportunity was Apollo 13, and Al was moving himself into the driver's seat in Cooper's place.

It was a power play, pure and simple, with the horses to back it: Al Shepard and Deke Slayton. Al was pushing for what he believed to be rightfully his, while Deke was eager to establish the precedent: a man wasn't eliminated from space flight *just* because he was older, or *just* because he had been grounded, or *just* because he had held a desk job for ten years. It was a popular cause with many on the NASA team. The public, predictably, was even more sympathetic to Al's side. It would do more for men in their late forties, emotionally, than all the Geritol and lunchtime jogging combined.

For those of us working on the Skylab program at the time, it was like listening to the noises of a distant crowd. For those in the arena, such as Gordon Cooper and Jim Lovell, their futures were at stake. Inevitably, the rest of us began to take sides. It became a matter of principle. If "they" could screw someone in this way and establish a precedent, "they" could screw any of us—which should have been news only to the most naïve.

It was obvious to anyone sensitive to office politics that Gordo was fighting for his life in the astronaut business. He had not flown a mission since 1965 and had served two successive assignments on a backup crew. He was being had, and he knew it. In that period when he had hoped to be in prime crew training, Cooper was flying a T-38 back and forth to Washington, trying to rally support at NASA head-quarters, trying to save *his* moon flight from the sentimental wave now building for Al Shepard.

It was a fruitless effort. During the years when Deke and Al had been shaping the organization and working with the people who counted, Gordo had pursued courses that generally cast him as anti-establishment. The chickens were coming home to roost; all the trips to Washington and all the appeals to the lords of the manor would be to no avail. So Gordon Cooper, the youngest of the Mercury astronauts, the one who had flown the best of the Mercury missions, and the one considered most likely to prevail in the contest to reach the moon, had closed out his space-flight career with Gemini 5—five years earlier.

Shepard had beaten back the challenge of Gordon Cooper in that apparent mismatch, but he still wasn't home free.

Jim Lovell had Apollo 14 sewed up, and the makeup of the 13 crew hardly seemed to affect him. But he was indignant that anyone, even his boss, could inject himself into the middle of the flight schedule without having served the same apprenticeship as those who were taking their turns in the rotation. We heard one day that he had threatened to resign over it, and we waited for the next round.

At this point, all the Shepard assignment lacked was approval from headquarters: specifically, the signature of George Mueller, the associate administrator. With time growing embarrassingly short for the crew assignment, Mueller made his decision. He didn't like the procedure any more than most of the astronauts did. There was no way that Shepard would be on Apollo 13. Mueller did not approve of the giant-leap-to-the-prime-crew tactic, and felt there was too little time for Al to reach the proficiency of prior crews.

Immediately after that, Deke Slayton approached Jim Lovell at the mission debriefing for Apollo 11, and asked, "Can your crew be ready for Thirteen?" Jim, who had just finished backing up Eleven, gave him the obvious answer, "Yes!"

Shepard still had to secure Mueller's approval for the mission for which he had been quietly training for at least six months. Through the efforts of Dr. Gilruth and his buddy, Deke, Al was able to get a headquarters OK for the Apollo 14 assignment. He had lost one and then won one, a fortunate sequence which was to insure the success of his efforts to land on the moon. His perseverance was exceeded only by the successful sixteen-year campaign of Deke Slayton. It was great for Al, but the waves rippling through the Astronaut Office left some long-term hard feelings.

Mine were not among them. It is to the credit of those who selected the Mercury Seven in 1959 that both these men, Deke and Al, were able to maintain that self-discipline and control which fulfills great ambitions.

Lovell was still living right. If Jim fell in a creek, he'd come up with a trout in his pocket. Those instincts would certainly come in handy before Apollo 13 was over. He was now going to become the first man to make four flights into space.

It was a fluke that Jack Swigert happened to be on board. It was to have been Ken Mattingly's mission. He had lived through all the work, problems, and excitement that go with a prime crew assignment. Then, three days before lift-off, it was learned that one of astronaut Charlie Duke's kids had come down with the measles, and Mattingly had been exposed on his last visit to Houston.

Ken was one of those rare birds who had never contracted measles as a child. When Chuck Berry and his medics found out, they threw themselves into the problem with abandon. Here was a live crisis at a crucial time, practically on the eve of a flight. Well, a potential crisis. A near crisis? By the time the doctors got through, we knew there'd be some kind of crisis.

The ruling was clean as a scalpel: Ken was grounded for the next ten days. The fallout was unreal. A disease little kids get had set the space program spinning. These were the options: the mission could be postponed a full month to meet the next launch "window" for the selected landing area on the moon; the backup crew could replace the prime crew and launch the mission on schedule; or Mattingly's backup, Jack Swigert, could move up in a one-for-one substitution and launch on schedule. They were fortunate to have a second talented Command Module pilot on the team.

Crew members had never before been replaced individually. More importantly Swigert had spent most of the preceding thirty days not training, but playing social secretary for the prime crew—the tedious but accepted chore of arranging rooms, invitations, and tours for their guests at the launch. Regardless of what was at stake, not one of us ever expected to see a crew broken up two days before launch. Some mixed-crew training was always done, but essentially we existed as a unit. The last six weeks before lift-off, the backup crew concentrated more on getting the prime crew ready than on anything else. So when the decision came, it was a stunner. After Thirteen, backup crew members recognized that any one of them could, at the last minute, be tapped to fly the mission.

Ken Mattingly was a meticulous guy, trained to a fine edge, and some questioned Jack Swigert's ability to step in and adequately replace him. Those of us who had worked with Jack had little doubt. He had worked with our team preparing for Apollo 7, had picked up hundreds of hours of simulator time in support of Apollo 10, and, finally, had served nearly a year on the backup crew for Thirteen. For all those credentials, and in spite of the pressure he faced, a demonstration of his proficiency was ordered by the top brass. It was to be a full-up simulation—a sort of *mano a mano* with Mission Control. In a twelve-hour ordeal his performance was flawless.

Jack Swigert was a real pro. I first met him when we were both going through astronaut selections in the summer of 1963. He was a friendly, appealing guy, big for an astronaut (a former college football lineman), with an unusual walk that was considered his trademark. His feet were so flat one expected a slapping noise when he walked, as if the soles of his shoes were loose. He walked along on those paddle feet as if he were behind a mule and a plow.

Jack was passed over in 1963, with indications being that he was deficient in both test-flying experience and academic background. He set out immediately to correct it. Jack enrolled at Rensselaer College and acquired a master's degree in aeronautical engineering. He then went on to fly the test program on the inflatable Regallo wing for Rockwell when that technique was being considered for incorporation into the Gemini program. With that to round out his credentials, Jack was selected the next time he applied. He brought to all of his activities the singlemindedness of a train running down a railroad track.

No flight was to be more hounded by irony than that of Apollo 13. Swigert was to jump in without any of the usual privileges of a mission, without time to even invite his friends and family to the Cape. He would be sent off to the moon in a Command Module that would explode before it got there, while Ken Mattingly brooded at home without ever developing the measles. It was a sad and unnecessary change creating near unanimous resentment from the Astronaut Office.

The third member of the crew, Fred Haise, certainly deserved more from his career than the unfinished mission he finally flew. Like Neil Armstrong, he had been a NASA test pilot before joining the astronaut corps. The year he reported to Houston he had won an award from the Society of Experimental Test Pilots for the best technical paper. In a business where the glamorous, high-speed jets usually stole the show, Fred's paper was on a test series he had conducted in civilian light aircraft. He was with NASA for a career, not just a leap in space. (That career looked like it would take a new direction a few years later when Fred crashed and was badly burned while ferrying a replica of a World War II Japanese aircraft for the Confederate Air Force.*) On Apollo 13, Fred was the resident LM expert, and he was scheduled to walk on the moon with Lovell.

When an oxygen tank exploded on Apollo 13, it plunged Mission Control, America, and the rest of the world into a general alarm. The planned mission was cancelled by acclamation, leaving one question: can they make it back alive? My personal confidence in the safe conclusion of the mission was based on the presence of Jack Swigert. Mission Control's performance during that period, the four days following the immediate crisis to splashdown, was one of its finest hours. But it was, in many respects, a flight-crew show.

Within fifteen minutes of the first report of the explosion in the Service Module—and only minutes after television's Jules Bergman had created a vivid impression of the crew expiring before making it back to earth—Dave Scott, Rusty Schweickart, and I were seated in the Mission Control viewing room Monday-morning-quarterbacking what was taking place. Bergman is one of the more knowledgeable science reporters on television, but still only about one-tenth as informed as he thinks he is. I believe his reporting on that particular evening bordered on the irresponsible.

When we arrived, the flight controllers were still huddling, going through the usual crisis format: What happened? Where are we? What are we going to do about it? As we monitored the air-to-ground loop, it was obvious that Lovell, Swigert, and Haise had taken the essential

---

* A flying air museum based in Harlingen, Texas.

steps even before those who were safe in Mission Control had time to call their first meeting. While the ground was debating whether to power up the undamaged Lunar Module—and if so, what systems— the crew, recognizing there was no alternative, was standing by for the official go-ahead on a decision they had already begun to implement.

The question of Thirteen's safe return was dependent on whether there was enough electricity, water, and oxygen remaining to sustain three men for the sixty-five hours required to reach splashdown in the Pacific Ocean. It was as simple as that. The flight crew's concern was with those actions necessary to stabilize and resolve the immediate problem. The degree of success was academic. They were like the mouse who no longer wanted the cheese, he just wanted to get his tail out of the trap.

From the moment of the explosion, the situation was deteriorating. Within a matter of minutes the flight crew was aware of what questions had to be answered to determine if they might die of old age or considerably sooner—possibly within hours. That was the moment for which their years of training had prepared them. They were attuned to it. Living in a spacecraft, moving through the environment it was designed to tame, one develops additional sensitivities. He becomes as aware of his supply of electricity, water, oxygen, and—especially— cabin pressure, as a businessman is of the weather when he rises in the morning, or of not cutting himself when he shaves, or of not walking in front of freight trains.

All of which points up the difference in outlook between the crew in space and the earthlings in Mission Control. Flight controllers are rarely in a position to contribute to time-critical decisions that may confront the astronauts as they whistle through the galaxy. What they can do, and do superbly, is monitor each of the instrumented points on the spacecraft (it would be impossible to display all of these to the crew). This information is fed to three of the largest IBM computers in captivity, manned by the most competent technicians who have at their fingertips the collective memory of all preceding missions. This is at once the strength and the weakness of the system. It takes time to digest the information available but this same time-lapse enables a precision that obviously can't be achieved with an immediate response.

Of course, the astronaut in trouble isn't looking for precision at first. At that moment he is worrying about what road to take, not how far the road will lead. This is not a reflection on Mission Control since many mission phases, and objectives, could not be accomplished without the ground controller's precise long-term data. The information available on board is never as complete and rarely as accurate.

The Apollo 13 crew performed the steps that would guarantee the security of their spacecraft and began powering up the Lunar Module to keep vital functions operating. It was a relatively long time, by their scale, before the ground was able to provide a confirmation of their actions, suggest some improvements, and confirm that they would, in fact, make it.

As Scott, Schweickart, and I studied the situation, it became apparent that the flight crew had done all they could. There remained only to trim it up a bit and wait for the trend information that would let them know just how much they had to tighten their belts. Ten minutes after we arrived I returned home and went to bed.

Out of those initial strategy sessions held at Mission Control, minutes after the explosion, came a detailed procedure for the Apollo 13 crew to follow to the safe conclusion of their aborted mission. It was designed to conserve the remaining resources and get them home in the shortest possible time. The crew wasn't exactly in Fat City, but they were alive and they were coming home. Thanks to the expertise of the boys in Mission Control, they were able to relax a bit.

The attention of the world was captured by this accident in space. The oxygen-tank explosion was an occurrence so unlikely that NASA had not even considered it a subject for contingency planning. The contingency plan which was born on the spot enabled the world to watch the crew of Apollo 13 splash down in the Pacific, closer to the carrier than any prior mission. It was a recovery from disaster that many regard as a major technical miracle.

Since the Thirteen crew had not been exposed to lunar dust, they were spared at least one burden: quarantine. The last crew scheduled for that treatment was Apollo 14 (the ground rule was three clean bills of health to warrant cancelling the treatment). Those of us who knew Al well awaited the explosion. Stick Al Shepard into isolation for twenty-one days? We had to see that.

We underestimated how badly Al wanted that mission. No more emotional battles for him. The czar of the office, he of the steely-eyed stare, had decided to concentrate on his training and tilt no windmills. He accepted the three weeks as part of the job. He didn't even whimper when they attached a three-week prelaunch quarantine as well. The Apollo 13 measles scare had taken its effect.

George Mueller had decided if Shepard would or would not be ready for Thirteen. But now Al was back in control of his own destiny, and he wouldn't be found lacking when the bolts blew on Apollo 14. In the years when he was flying very little (because of the second-pilot restriction), he remained a fine aviator, a prime example of my theory that being a good aviator is more a state of mind than a learned manual skill. That same ability, exercised several steps further, is essential to prepare for a space flight. Al went at it with a vengeance, including his physical conditioning program. By the end of the mission he was right up there with George Blanda and Lawrence Welk as a hero of the Geritol set.

Redheaded Stu Roosa, the Command Module pilot, was an air force major when he flew to the moon with Al Shepard. I say "with," because Apollo 14 was Shepard's mission. America gave it to him like a necklace in the same way that Apollo 7 was always known as Wally Schirra's mission.

Stu was an avid hunter and a great bird shooter who had a boyishness and an earnestness about him that would do justice to Huck Finn. He tolerated the occasional carousing of his friends, but he was a straight arrow.

Ed Mitchell was the third member of the crew, a navy officer, bright, articulate, and an independent thinker in a place and time where independent thinkers were not really in demand. If it had not been "Shepard's flight," the news media would have gotten all the copy it craved from Ed Mitchell, who was into ESP and psychic phenomena.

By the time Apollo 14 was ready to roll, Ed had suffered his share of frustration and was fed up with many parts of the NASA system. He had other goals to pursue and was intent on reordering his life—which included a divorce. Several months before the flight, he let it be known that he would be leaving the astronaut corps shortly after the mission was completed.

That posed no problem for Deke; he simply made it plain that he had planned for Ed to back up Apollo 16. If Ed couldn't wait around that long then someone else who could would be given the opportunity to fly Fourteen in his place. Ed stayed for both assignments. No one thought less of him for changing his mind. We all had a commitment to our dreams.

The last four missions, beginning with Apollo 14, were technically a string of near-monotonous successes—beautiful jobs on difficult missions that few would remember were it not for some relatively unimportant fringe developments. The one that attracted the most notice, unfortunately, tended to characterize this two-year period in manned space flight.

The most complete scientific explorations of the Apollo program were carried out by the crews of Apollo 15 and 16. Yet Fifteen is best remembered (when it is at all) as the crew that carried envelopes to the moon. It was the dog days for the space program and a tragedy for the Apollo 15 crew: Dave Scott and the two rookies, Al Worden and Jim Irwin.

Dave was a three-time veteran who, on many occasions, acted as a self-appointed conscience of our group in the face of commercial temptations. He often advised restraint when so-called "good deals" surfaced, though his counsel was frequently ignored. So it was with bitter irony that so much adverse publicity for NASA was generated, so many careers shaken, and so much of the carefully orchestrated astronaut image blown on the fallout of Dave's mission. To many Americans, stamps and postal covers (envelopes) are a national pastime, but to serious collectors and dealers they are big money. Scott had been an amateur philatelist, like many of us, and may not have realized that with the Apollo 15 caper he was turning pro in a big way.

After the postflight tributes died down, the three of them settled in for another year of serious work as the backup crew for Apollo 17, the last of the lunar landings. At the completion of the program, Worden and Irwin planned to resign at their own convenience and move on to other careers. Scott could have been expected to do the same, unless he wound up running the Astronaut Office or some other part of the Manned Spacecraft Center. Instead, at the height of their professional achievement their futures turned sour.

In the fall of 1971, NASA became aware that a brisk philatelic business was being conducted in Apollo 15 first-day covers that had traveled to the moon. By the spring of 1972, they had completed an internal investigation and initiated disciplinary action. And by June or July of that year the news media were getting their teeth into it and, in the name of the public's "right to know," were publishing details of a full-blown scandal.

The picture that slowly emerged was of a crew that had carried approximately 630 special first-day covers to the surface of the moon on board the Lunar Module Falcon, most of them unauthorized. They had returned with the crew and carried postmarks at several critical stages of the mission, including on board the recovery carrier *Okinawa*. During the long return flight from Hawaii to Houston, the three of them had signed each of the covers, further enhancing their value. When they were able to catch up on neglected business, one hundred of these covers were shipped off to a German stamp dealer where they quickly sold out at $1,500 apiece.

NASA's first comment came from John Donnelly, associate administrator for public affairs, to the effect that there was no profit motive on the part of the crew and therefore the moon trips were unsullied.

When asked why, Donnelly replied, "Why Dave just wouldn't do a thing like that," reflecting the high regard and simon-pure reputation of Scott at that time.

The next release stated that the three were to have split $21,000 as a sort of "transportation fee" which was earmarked as trust funds for their children's education. Recognizing that apparent impropriety, they chose not to accept the money and did not profit from the deal.

What of the other 530 envelopes? Well, 88 were really Apollo 12 covers belonging to Barbara Gordon, Dick's wife. The balance of 440 were apparently to remain with Scott, Worden, and Irwin. At $1,500 apiece they would have been worth a cool $660,000.

If any of us realized how large and lucrative the philatelic business really was, it was Barbara Gordon. But her efforts also reflected the kind of harmless nuisance most of us took stamp collecting to be.

During the Gemini program, Barbara set out to collect for each of her six children a complete set of first-day covers marking each of the space missions. When autographed by each of the crewmen, the collection would not only serve as an attractive souvenir, but should also appreciate handsomely in value over the years—a thought that no doubt occurred to Barbara.

I had to admire her tenacity in persuading her husband's friends to keep signing stacks of commemorative covers from their flights. Lo and I lived just one block from the Gordons, and periodically, I could expect a call from Barbara asking if I would be home for a few minutes. A few moments later she'd pedal up on her bicycle with her latest Apollo 7 covers to be signed. Her biggest haul came after Dick's last flight, on Apollo 12, when she circulated more than a hundred first-day covers for signing by Dick's crewmates.

I have no idea what the going rate is for first-day covers, stamps, postcards, and other such items, but I doubt if Barbara Gordon missed a set in over six years. With what she collected, and her potential for trading, it may represent the most complete and valuable collection of space items in the postal universe.

My first reaction when the scandal leaked out was shock, not just that the incident had occurred, but also as to who was involved. It did shed some light on a memo that Alan Shepard had sent to all astronauts only a month earlier warning us to be alert for the commercialization of various articles which we autographed almost daily for collectors around the world. Enclosed was a one-page newsletter from a big collector and small-time dealer listing some of his wares for sale. A few examples:

   1. Agnew, Spiro—Vice President—signed card $10.00
   2. Armstrong, Neil—Historic photo of the Wright Brothers' first flight signed by Armstrong . . . photo is a NASA print $50.00
   3. Allen, Joseph—Scientist Astronaut—very scarce FLOWN COVER carried on a T-38 by Allen on a training mission . . . this is one of a few rare flown covers to be offered on this list $50.00

   (*Flown?* Houston to Cape Kennedy in an airplane!)

I could feel sorry for the entire crew, but my heart went out to Dave. He had always set high standards, not only for himself but for all of us.

If he failed one had to wonder if it were really possible for virtue to triumph, ever. Dave is the kind of guy who will carry the shadow on his career like an open wound.

Dave and I had been getting along better for several years. Almost from our first meeting I had had mixed feelings about him. He in turn seemed to resent my whole approach to the job. I admired his talents and dedication, but he had what I took to be a holier-than-thou attitude. We had a personality clash that reached a climax of sorts the day of the Phantom Buzz job. It was an incident that reveals a bit about both of us.

I was returning to Houston from Patrick Air Force Base one Saturday morning, near the end of the training grind for Apollo 7. It had been a hard week. The sky was clear and seductive. I was making good time and the adrenalin was flowing.

It is a biological fact that the more an aviator is enjoying his flying, the more susceptible he is to the temptation of flat-hatting. One hundred miles out, I dropped the nose of my T-38 and pointed it toward the community of Nassau Bay, where many of us lived. Over my house (and a block farther, Dave's) I was fast and low—about 100 feet—and pulling up the nose I hit both after-burners. It was like a bomb going off. At 5,000 feet I immelmanned out and swung quickly into my approach to land at Ellington Field.

In less than twenty minutes I was walking through my own doorway, but the phone was already ringing. Stu Roosa had landed a little ahead of me and was calling from the hangar to warn me that complaints were coming in, and I was thought to be the culprit. On Monday morning, Deke summoned me for the traditional commanding-officer-to-young-lieutenant-lecture ("Knock off that childish bullshit") and sent me away to sin no more.

The topper came when Dave Scott, obviously still burning, came storming into my office.

"Walt," he demanded, "were you flying at eight o'clock Saturday morning?"

I frowned, as if trying to remember that far back. "Yes, come to think of it. I flew in from the Cape."

"Was that you who buzzed Nassau Bay?"

Sheepish grin.

"Dammit, you scared hell out of Lurton [his wife]. If I ever catch you doing that again"—he glared at me—"I'll . . . I'll have your wings yanked."

For once in my life I was speechless. It wasn't just that Dave was a hot-rod jet jockey himself and should have known better. It was his attempt to throw around weight that he didn't have. I mean, he wasn't a general yet. It was like threatening to make a citizen's arrest because some guy had parked in your space. (Esthetically, it was a pretty good buzz job, but it was the last one I ever made . . . in Houston.)

Our differences were largely the result of poor chemistry, which seemed to straighten itself out during the Apollo program. I could take no pleasure from Dave's misery in the envelope case. For one thing, quite a few of us would get our cuffs muddy before the dust settled.

Dave was younger than each of his crewmates but in a sense was the "old man" of the bunch. Besides being the commander, he had already been through the routine twice and was expected to keep the two rookies from making mistakes, both technical and ethical. Instead, there are some indications, as well as insinuations by Worden and Irwin, that they were following Scott's lead.

Deke had few alternatives and no reason to hesitate. In 1972 he had astronauts coming out of his ears and damn few missions to fly. He could afford to purge all three and did. They were relieved of their duties as the Seventeen backup crew, and reassigned or nonassigned. Jim Irwin moved his planned retirement forward six months and went out to preach the gospel with his own Operation High Flight, an organization incorporated about the time of Apollo 15. Here was a man truly mission-oriented. His message, "I felt God's presence with me on the moon," provided a tremendous boost to what would otherwise have been just another lay preacher's effort. After the censure, Irwin rebounded by weaving the moral lesson of the first-day cover temptation into his sermons. He also admitted publicly that NASA had no alternative. High Flight is still going strong.

Worden was on a geology field trip in Arizona when he learned that Deke had arranged for the air force, his parent service, to recall him. He had been waiting for the other shoe to drop, but this move stunned him. He had little enthusiasm for leaving NASA or playing Wild Blue Yonder again, at least until he had completed his twenty years' federal service (for retirement) as an astronaut. How are you ever going to get them back on the farm, after they have seen the lights of Cape Kennedy?

Worden decided to fight. I heard he called a congressman or two and then went over the heads of Deke and Chris Kraft to argue his case with Dale Myers, the deputy administrator for manned spaceflight. Myers gave him a reprieve, long enough to find a slot at NASA's Ames Research Center to finish out his twenty years. This was also the time when Worden was reorganizing his own life and getting a divorce so it suited him just fine.

The events seemed to bewilder Dave for months. He nearly dropped from sight but when we were together he was more subdued and congenial than I had ever known him. Dave was shuttled into a desk job at the Manned Spacecraft Center and a year later moved on to the NASA Flight Research Center at Edwards Air Force Base, California, where he eventually became the administrator in charge. Strong and capable, he has bounced back.

Scott's actions in the envelope case I can only attribute to a lapse, one that, no doubt, Dave doesn't understand himself. He was in for the duration and he brought pride and ambition to his work. Therefore, the consequences of his folly were far worse than for the rest of us.

Apollo 15 may have been the first crew to commercialize on first-day covers, but it was not the first to carry them. They were carried on every flight since Apollo 11, and the practice may have started earlier. From Twelve through Sixteen they had been obtained through Al Bishop, then working for the Howard Hughes organization. They were limited orders of a few hundred, featuring the crew emblems in the left corner and were the nearest thing to an "official" cover. The crews carried only a few and usually returned a couple to Al, signed, with gratitude. To my knowledge, Bishop has never sold a one and never made a dime off his relationship. He was simply a fan.

To Al, Apollo 15 was no different from any other flight, except for a phone call he received from Hal Collins, the Astronaut Office manager at the Cape. I believe Collins told Al Bishop that the crew would like to know whether he could obtain some very *lightweight* envelopes for them. Al said that he'd be happy to do so. He was unaware then that many of them would be smuggled on board the next lunar flight.

Al was trusted. That's why many of us imposed on him with our problems, special requests, and sometimes matters which we would rather not share with NASA, like my doctor's appointment four days before launch. In that respect, we often took advantage of him. This time Al was badly used and he emerged as a scapegoat.

Hal said that they wanted to print twelve hundred of them, and asked whether Al could handle that.

Bishop told him that this would present no problem.

After the flight, when it became apparent that the deal had backfired, Bishop got another call from Collins. He began by warning Al that a big stink had developed over the envelopes.

When Al asked him what he was talking about, Collins told him that he would let Dave explain it to him.

Dave never called. No member of the Fifteen crew did. But Bishop's name got dragged through a Senate hearing as though he were the Mr. Big of an international stamp conspiracy. It made headlines in Las Vegas, where he lived, and eventually cost him a good job. His taxes were audited and most of his astronaut "pals" dropped him like a load of barnyard manure. It took Bishop several years to recover from his helpfulness.

I resented the news media's implication that the near-perfect job performed by the Apollo 15 crew was any less an accomplishment, just as I had resented the implication that we were all affected by Buzz Aldrin's public acknowledgment of a nervous breakdown.

It would be nice to say the rest of the astros were all indignant, but the truth is many of us were living in glass houses. During the course of the investigation other "lapses of judgment" came to light. One that compromised most of us was a series of "business deals" with the same middleman that proposed the moon cover caper to Scott's crew.

My own involvement began with a 1969 phone call from Walter Eiermann, whom many of us had met socially in Cocoa Beach. Eiermann said he was negotiating with many of the astronauts to sign 500 plate blocks of stamps issued to commemorate the individual's flight for a payment of $2,500. My answer was probably common: Yes, I'd be interested if others in the group were doing the same thing. The money was attractive and it wasn't explicitly against any regulation, but I was *not* interested in becoming a trail blazer.

Three months later he called with a list of those who had already committed. His goal was to sign every man who had flown an Apollo mission, and he seemed well on his way.

I suppose we could argue that the money went for a good cause. Most of it went for living expenses of families accustomed to the added income from the *Life* contracts, which had dwindled to nearly nothing. Some of it went to charity at the time, and I know some more did after the story broke. The investigation by NASA and the justice department may have caused a few amended income tax returns as well.

Through the Fourteen crew, the only Apollo astros I was ever certain did not take the bait were Bill Anders, John Young, Mike Collins, Buzz Aldrin, Fred Haise, and Alan Shepard. There may have been others. Among those of us involved, the only one I know of who ever received any obvious disciplinary action was Jack Swigert, and then it was not for signing (which apparently was technically legal).

Jack's problem was that when NASA investigated the whole affair, he at first denied having signed the stamps. He did so to protect his assignment on the U.S.–Soviet mission—and it ended up costing him that very prize. He had already been selected for the crew, along with Stafford and Slayton, and was happily enrolled in Russian language courses. There only remained the official public announcement.

A few weeks later Jack had a change of heart or conscience, or maybe someone advised him to clear the air. He went to George Low, the acting administrator of NASA, and admitted that he had signed but said he'd given the money to charity. Low's decision to remove Jack from the mission was based on the fact that Jack had lied. Needless to say, Jack was shattered.

Was it illegal to sign the stamps for profit? No. Was it morally wrong? Maybe. Was it a mistake? Hell, yes! At times it still has the echo of a bad joke, but it added to the growing media picture of heroes with feet of clay. Yet, with all the times we could have exploited our careers on the wild deals that held the lure of big money, the stamp signings seemed a relatively innocent and safe way to earn a modest sum.

Our image wasn't enhanced any by other things that were brought to light on the periphery of the Apollo 15 investigation. The Franklin Mint got a grubby hand involved in the Apollo 14 mission after being partially successful on Thirteen. The mint wanted to commercialize on medals that had flown to the moon. Getting them on board the spacecraft was apparently not too difficult; all of the crews to date have commissioned and purchased, exclusively for their own use, commemorative medals, some of which were carried on the flight.

Prior to the Apollo 13 flight, the crew was approached by (once again) a middleman with an offer of several hundred free sterling silver medals in exchange for returning fifty of them, which had been to the moon, to the Franklin Mint. The mint would melt down these fifty and coin 130,000 medals that had "been to the moon." When Thirteen aborted without landing on the moon, Lovell returned the single medal he had carried in his flight suit which caused the scheme to fall through. Later, Shepard and the Fourteen crew agreed to more or less the same deal. The scuttlebutt around the office was that Apollo 14 carried a personal package on board weighing forty-two pounds.

Both Lovell and Shepard attempted to hold down the exploitation, but with little leverage they were unsuccessful.

Still another flap involved a now-famous sculpture called, appropriately, "The Fallen Astronaut." Dave Scott and Jim Irwin, with the purest of intentions, placed it in a small crater on the moon on August 2, 1971, together with a plaque listing the names of the fourteen astronauts and cosmonauts who had died in the effort to explore space. The tiny aluminum statue now seems a symbol not only of those men but also of the crew that carried it there.

The sculptor flew to Florida two months before the flight to offer his sculpted figure to Scott and his crew as a memorial to those who had perished. NASA's approval was obtained with the understanding that there would be no commercial exploitation.

At the postflight conference, the crew disclosed the project and Dave made public a photograph of the plaque and statue—without identifying the sculptor. Three months later, the sculptor, who has been doing space themes for twenty years and now resented being ignored, broke his silence by offering 950 copies of the statue to the public at $750 each as "exact replicas" of the first work of art on the moon.

In no way did any astronaut profit, nor was it ever meant that they should. Of Scott, the gallery director said, "He was a shit about some things, but he was the one who got the goddamn thing on the moon. He was sincere, but nervous about the publicity. He was the one who really wanted to be a general."

So it didn't begin with the first-day covers; they were just the key with which the dirt was dug up. Sometimes we were willing accomplices (as in the case of the two watches of a major manufacturer which were "evaluated" on Apollo 15), but more frequently we were naïve or just plain outsmarted.

Coincident with the flap over the astronauts' various postal enterprises, Apollo 16 flew and became the most forgotten of all the missions. Probably not one in a hundred can name any of the three crew members, and yet in terms of execution—and scientific returns—it has to rate along with Apollo 15 as one of the most productive. No scandals, either. The flight was commanded by well-trained, super-trained, always-training John Young, who (glory be) hadn't even signed any stamps.

Apollo 16 was a clear demonstration of the waning interest in space (although Apollo 17, as the final act, would inspire a brief sentimental revival). The clues had been coming for some time: less congressional support, budget cuts, reduced TV coverage, smaller crowds at the Cape, and to me, at least, a change of attitude in the make-up of those crowds. The earliest launches were viewed mainly by those involved in the program or on the fringe of it or by diehard space buffs. Toward the end we were drawing more tourists, more families, and there was more of a county fair feeling. Cocoa Beach was in a state of transition, with the people, the merchants, and the money moving from the Fantasia of space to the reality of Disney World, 40 miles down the road at Orlando.

So with a minimum of pomp and fanfare, John Young, Ken Mattingly (the only man to toil through two consecutive prime crew assignments), and Charlie Duke went to the moon aboard Sixteen— and the Apollo program was down to one.

Apollo 17 would be described, repeatedly, as "the end of the beginning." It would break new ground in at least three respects. As the first night-time manned launch, visible for 500 miles in any direction from the Cape. For the first time a scientist-astronaut, Jack Schmitt, would fly in space. Finally, tight security was instituted around the astronaut families throughout the final training period.

That latter item reflected a frightening new development in world politics: the age of terrorism. Not long after the slaughter of the Israeli athletes at the Munich Olympics and rampant terrorist acts elsewhere, NASA wisely reappraised the security for its last Apollo launch. It was the kind of theater that could easily have tempted such madmen.

Gene Cernan, the spacecraft commander, and Schmitt were joined by Ron Evans, the middleman in the spacecraft, who had worked on our Apollo 7 support team and four years later was finally getting his own flight.

There was a feeling about Apollo 17 of a graduation day for the class that had enrolled, fourteen of us, nine years before, and a recognition of the last hurrah for manned lunar flights.

Seventeen had a beautiful midnight sendoff. Mingling with the crowd, as an alumnus, I recorded the final count. You can hear it over the hum of the crowd noise, the monotone voice, no more emotional than the public address system at an airport telling you that Flight 52 to LaGuardia is ready for boarding. But my scalp tingled. I heard my own voice whispering, "Come on Seventeen, let's go," and it didn't sound like me. At lift-off the roar of the crowd was spontaneous—pure excitement. You could feel the vibrations. And the sound rolled over us, even as they hit us. Dammit, people cried. How do you tell anyone that you saw the earth tremble?

On their return from a good mission, Cernan, Evans, and their wives, along with the bachelor Schmitt, made the last grand astronaut tour. Gene Cernan was like a politician at a Fourth of July barbecue. He had his speech down pat, a sweeping discourse that touched all the bases: pride, humility, patriotism, the world view, family, and a heart-felt thanks to anyone who had anything to do with the space program. It was a speech that didn't make you think very hard, but it made you feel good. Gene was quick on his feet. He reminded people of those forgettable last words uttered on the moon just as their Lunar Module was lifting off, "Okay, let's get this mother out of here!"

Jack once complained, tongue-in-cheek, that when he would inject something new or clever in his own brief talk, "the next day all the good parts would wind up in Gene's speech, and I'd be left with nothing to add."

Ron Evans became the comic relief who would do an entire night club monologue, with gestures, on the astronauts' best-kept secret: "How to pee-pee and poo-poo in space." If it sounds coarse, it was, and in five fast minutes Ron provided the audience with an empty, if funny, view of a man in whom the government had invested $5 million. It was an embarrassing performance to many of his friends, but Ron was the hit of the show wherever they went.

Jack Schmitt was a quiet, private fellow who never appeared totally at ease with his teammates. He didn't relish the public relations aspects of the postflight scene, and easy camaraderie wasn't his bag either. As the first scientist-astronaut to fly in space, Jack Schmitt held his own, accomplishing no less (and no more) than anyone else. If the lunar scientists expected Jack to carry the flag for them, they were wrong for at least two reasons. One is the rather narrow view held by the scientific community that most of the astronauts were nothing but "dumb fighter jocks." Secondly, they failed to recognize that the success of a mission was controlled more by the operational realities than by the particular skills and serendipity of an individual crewman. Add to this the hard facts of life of operating in a pressurized suit and the environmental constraints, and one can readily see that the real triumph is in just being there and moving around.

Schmitt's scientific training did allow him to do one thing which no one had done before. On one of their moon walks he came upon some orange rocks and was elated. Eureka! Jack had found the moon's youngest rocks—or so it seemed at the time. He proudly proclaimed their age as "less than twenty-five million years." After the rocks and soil had been analyzed at home, it was determined that Jack missed in his on-the-spot observation by a mere three and a half *billion* years. I dare say a "dumb fighter jock" would not have ventured such an ad hoc assessment.

That mistake didn't affect the fervor and eloquence with which he expressed his feelings on space flight, as he did when the Apollo 17 crew addressed a joint session of Congress.

"I would like first to tell you about a place I have seen in the solar system," he began, "the Valley of Taurus-Littrow. . . .

"The Valley has been unchanged by being a name on a distant planet, while change has governed the men who named it. The Valley has been less altered by being explored than they who have been the explorers. The Valley has been less affected by all we have done than have been the millions who, for a moment, were aware of its towering walls, its visitors, and then its silence."

It was a magnificent way to describe the completion of the greatest technological achievement in history.

As I look back over the years of Apollo, there were a few surprises in our third group. That Dave Scott would fly three missions startled no one, but that he would share that distinction with Gene Cernan came as a distinct upset. Gene ranked as one of the two miscalculations in my 1964 peer ratings. As events unfolded, the first seven of our fourteen went into space in this order: Scott, Cernan, Collins, Gordon, Aldrin, Eisele, and Anders. That line-up, of course, reflected the changes caused by the See–Bassett accident and the Apollo 1 fire.

Were it not for those fateful accidents, the order of flight would have been: Scott, Bassett, Collins, Gordon, Cernan, Chaffee, and Eisele. Which is precisely the order in which they appeared, relative to each other, in my peer ratings.

The most glaring discrepancy between my ratings, and the order of baptism, came at the top. Alan Bean was ranked Number 1, but he flew tenth—and last. Possibly friendship got in the way of judgment, but I prefer to think that later events on Skylab proved my assessment to be a better appraisal of his ability than those who originally controlled the assignments.

The Apollo program was over. The midnight launch that December 7, 1972 was the last time that most of the astronauts would be together to celebrate the years in which we rode that great flame into the heavens. We had competed, we had argued, we had often let ourselves down, and we had never gotten there with that poet everyone kept wishing we could take along to help us find another word besides "fantastic." But there were moments of undiluted joy and good company, and, finally, we did give science a shot at the moon. It was, after all, only the end of the beginning.

14

The Hyphenated Astronaut
Comes of Age

THE REAL CLASH in our space world, though it existed largely behind the scenes, was between the pilots and the scientists—the goals of technology against those of science.

Take Skylab. In the beginning, NASA assigned fresh astronauts to Skylab in much the same spirit the early Russians threw one another off the back of the sleigh to slow down the wolves.

To the surprise of no one, thirty-two American fighter pilots traveled into space before the first scientist. The string was broken by scientist-astronaut Jack Schmitt on the last Apollo mission. Of the next seven to fly, three were trained scientists before they became pilots. It was the beginning of a necessary transition from career aviators, trained to do whatever was necessary on the mission, to scientists, doctors, or other specialty passenger types who could wring the most out of a given area of investigation once they were carried aloft.

Scientist-astronauts were brought into the program as far back as 1965. It was clear even then that they would outnumber the aviators some time in the future. The arrival of those first six scientist-astronauts created a stir around the Astronaut Office. This was a new kind of animal to be integrated into the system, and the older heads studied it warily. They quickly decided that the new breed was inferior, and worried that Congress might not know the difference, or care. Contact around the office was cordial enough, but it was clear that the scientists

244

had bigger handicaps to overcome than just the usual new-guy-on-the-block syndrome. They clashed with and even contradicted our public image, which we enjoyed, as the John Waynes of the space frontier. No milquetoast academic types would fit into that picture.

My own reaction to their arrival was relief, for virtually the same reasons the others were distressed. Until they arrived, the closest thing to resident scientists in the Astronaut Office were Rusty Schweickart and myself, both having reported to NASA as civilians straight from the campus. I can't speak for Rusty, but I worried about being a part of the team; I mean the varsity and not some kind of token to the scientific community. I was more than willing to exploit my academic background during the selection process, but once in, the thought uppermost in my mind was to remove any question about my flying credentials.

By 1965 the argument between the National Academy of Sciences and the Astronaut Office (notably Deke Slayton and Alan Shepard) had grown quite heated on the subject of what qualifications best suited a man for a trip into space. Deke maintained that a highly experimental flight test program like the development of manned spacecraft required specialists who had dedicated their lives to the operation of the closest related kind of equipment. The program could not afford the luxury of anyone who couldn't carry his own weight in such a flight test program. Furthermore, Deke's boys could handle just about anything that would interest the scientific lobby, and get the spacecraft back to earth as well. To fill a technical job, one sent the best technicians available.

Deke wasn't alone in that opinion. The rest of us were cheering like mad—this was no time for unselfishness. The National Academy of Sciences was just as provincial as we were. Space was such an appealing and unexploited area of investigation that the subject simply precluded objectivity. The Academy's position seemed to be that anyone with a yen for adventure could be a pilot, but only God could make a scientist.

There was no way, of course, that the scientists were going to be shut out of the space odyssey. They had a big hammer. At a time when the NASA budget was headed toward $6 billion annually, the agency needed all the friends it could find, particularly in Congress, which held a generous respect for science along with the other things it didn't understand. When the flak began to fly about spending $25 billion on a "moondoggle," NASA took the prudent course and caved in. Deke had lost the battle.

Slayton wasn't acting out of prejudice. He was simply the one closest to the heat and no one knew any better what the crew requirements would be for the next ten years. NASA was projecting the first lunar landing for late 1967 or early 1968, with repeat trips every two months thereafter, and was hoping the first Skylab would fly in 1968.

It would have been politic for Deke to man up on the basis of those projections, but as always he took the most realistic view. He also had, I think, the best appreciation of the kind of man needed to perform the job he would be expecting. It must have struck him as a waste of time, money, and talent to recruit scientists into the program so early, knowing at least five years would pass before any of them would fly. (That forecast turned out to be painfully conservative. Those who hung in there from that first group took their rides in their seventh, eighth, and ninth years of training. It was almost like being a career cadet.)

The scientist selections took place in an atmosphere that must have resembled a jury trial, with the defense and the prosecutors each applying their own standards to the jurors. The National Academy insisted on screening the applicants down to the final round, at which point NASA could apply its criteria. The Academy proposed to solicit the scientist candidates, when all those chosen in the past had aggressively sought the job, but Deke was able to insist that the newest candidates begin their careers in the same way.

The Academy gave little attention to a nominee's prospects for passing the NASA screening, concentrating instead on the totems of scientific manhood, such as a Ph.D., numbers of papers published, and years of research. Good prospects who lacked the purebred scientific pedigree were disqualified early.

When it came NASA's turn, what remained to be judged were each candidate's qualifications to meet the requirements of his new discipline and his motivation. Additional screening assessed their ability to measure up to the years of screening and filtering that each of the pilot astronauts had endured even before applying. One experience, for example, was to expose their zest, or aversion, to the job ahead. That was the one-hour, aerobatic flight in a T-38, which consisted of spins, loops, dives, and other fun maneuvers. That alone eliminated some candidates—I mean just the idea of it, not the actual performance.

When the first selection was over, NASA could accept only half the number of scientists it had hoped to bring into the program, which left precious little margin for further attrition. In retrospect, the internal politics and the outside pressure acted as a dose of bitter medicine that was good for what ailed us. The outcome was surprising and, in keeping with the other NASA successes of the sixties, the result of both a lot of effort and a little luck. We acquired the services of five very atypical scientists out of the six who reported that year: a medical doctor, a geologist, and three physical scientists.

There was a nucleus of aviator types. Curt Michael had been an air force pilot before earning his Ph.D. Joe Kerwin was a navy flight surgeon as well as a naval aviator. Owen Garriott, a university professor with no military experience, had still logged about 400 hours of civilian flying time. Ed Gibson was brash and bright; a hard-charging youngster who, in many ways, reminded me of a Roger Chaffee with no flight experience. Jack Schmitt was a nonflying specialist in lunar geology and the only one with a good case for inclusion in the Apollo program.

The last of the six was the unfortunate Dr. Duane Graveline, whose domestic problems forced his resignation even before flight school. His wife made it a "NASA or me" issue, and in the divorce filing that followed she accused him of "an uncontrollable temper." That was hardly appropriate for an astronaut, and the good doctor was asked to return home.

So then there were five, who might as well have spent the next seven years in a vault for all of the space exploring they got to do. Nevertheless, the felony was compounded in 1967 when NASA was again pressured into admitting another group of scientist-astronauts (eleven this time), whose scientific specialties once more ranged from geologist to medical doctor. They were collectively almost devoid of operational experience, and some appeared to have no real interest in acquiring it.

This was the group, quick on the uptake, that identified its own future by dubbing itself with the fine, military-sounding name of XS-11 (the excess eleven). Some of them would contribute in other areas, but by 1977 none had flown in space, and it seemed likely that few of them ever would.

One who seemed more than willing to play the astronaut game was Storey Musgrave. His background was a little unusual even in that fast company. Storey was a high school dropout who enlisted in the Marine Corps and four years later returned to get serious about it all. Before joining NASA, he had managed to accumulate the following academic credentials:

> Bachelor of Science in statistics
> Bachelor of Arts in chemistry
> Master of Business Administration in operations
>   analysis and computer programming
> Doctor of Medicine
> Master of Science in biophysics

In addition, he was working on a doctorate in physiology. In his spare time he had found time to accumulate 1,500 pilot hours and 200 free-fall parachute jumps. He also started a family of five children—all before his thirty-second birthday.

For him, air force pilot training was a ball. On the side, he was giving flying lessons and commuting to Denver on every free weekend to plug away at an internship in surgery. After pilot training the problem was not how to keep him from playing scientist, but how to keep him out of an airplane cockpit long enough to get some work accomplished.

At the other end of the spectrum was another medical doctor, Don Holmquist. Making a king-sized judgmental error, he threw in with the wrong side of a power struggle that had been going on for ten years. Holmquist was bright, young, and very immature. From the beginning I watched him curry favor with the NASA medical directorate, assuming they would be an asset in his corner at crew assignment time. Little could he know that that was like waving a red flag in front of a bull. He never did figure out what the Astronaut Office expected of their medical astronauts, and he was eventually asked to leave.

The senior doctor in the group was Joe Kerwin, whose effect on the troops was the other extreme from that of Holmquist. From the beginning Joe set out to be accepted as an unhyphenated astronaut. He treated his medical background as I had the study of physics: as an adjunct, a tool to be exploited within the broad guidelines of a new profession. His medical skills had to suffer while he concentrated on his new career, but many of us looked upon him as more our "personal" physician than any of the NASA doctors. There were damn few astronauts who didn't hide medical problems from our NASA flight surgeons, but seldom was anyone reluctant to discuss those problems with Joe. Judged by the respect they paid to the confidential doctor-patient relationship, the doctors at NASA with the highest professional ethics were those within the Astronaut Office.

Dr. Kerwin was always willing to make a house call for a sick child, even if NASA doctors were available. I'll never forget calling Joe one night when Brian had a fever of 104° and we couldn't locate the NASA doctor on duty. Joe lived only two blocks away and he rushed right over. We both got a real laugh when he pulled a rusty stethoscope from his bag. Joe didn't get much time for doctoring.

There were a few others with scientific credentials just as impressive as the rest, but who were also as enthusiastic about learning the profession of astronaut as any of the pilots. Some of the scientists didn't make it, and it didn't take long after their arrival to spot those having trouble making the adjustment.

The most famous scientist-astronaut dropout was Brian O'Leary, who gave as his reason for quitting, "Flying isn't my cup of tea." True, he was afraid of flying, but he was also out of step even with his associates. In the six months he spent in our office he had not learned what motivated us, but it was apparently long enough to convince him he did not like the whole setup, including the city of Houston. When he found out pilot training scared him to death, his mind was made up. He left the impression, "Well, I just wanted to see if I could get selected, and I did." No motivation.

O'Leary's case has to go down as a failure of the selection process. The selection board, from Deke Slayton to the psychiatrists, may not have appreciated the difference in motivation between the John Waynes who volunteered to kick the program off and the scientists who were recruited ten years later. It was much more difficult to maintain a good batting average when dealing with the raw, unfiltered bunch that constituted the scientist applicants. Pilot-astronauts had been successfully passed through many filters before their first application for astronaut training.

The guy who got by the longest before the astronaut equivalent of firing was Dr. Curt Michel, who left NASA to become chairman of the Space Science Department at Rice University. He had little more desire for the job than Brian O'Leary and no more understanding of how to fit in (if he ever wanted to) than Don Holmquist. He had applied for the program partly at the urging of friends in the academic community who felt his air force flying experience, along with his fine academic record, would insure success. Among the group approved by the National Academy of Sciences he looked like a good bet. After reporting in 1965, he was the most conscientious in maintaining a schedule of one day a week at the scientific laboratories of Rice University. In that respect he was dedicated to his real vocation, but it hardly improved his cynical and casual approach to the astronaut job. In 1969, after he took a year's leave of absence to devote full time to his scientific pursuits, it was quietly turned into a permanent arrangement.

When I was asked about the space program by scientist friends in those days, I tried to be honest. "It's the most exciting thing going today," I said. "If you want to be a part of it, then go ahead and apply. But if what you really want is to do 'good science,' then you'll be better off investing your time elsewhere."

That was the sadness, the failure, of the role designed for these men. The skills of some went unused while they underwent training that would be years in paying off, if ever.

The box score reads: seventeen scientist-astronauts brought into the program in a span of two years. Three quit before the end of their first year and two more were dropped as soon as it was convenient. The dozen who remained toughed it out over a course that had to be more frustrating than the path followed by the rest of us. Deke had been right when he said it was premature to recruit that many scientists into astronaut training.

Some hoped to beat the system by transforming themselves, through attitudes and actions, into *aviators.* Several graduated first in their pilot training classes, even though they were looked upon by the other cadets as old men. Those who endured did so knowing full well the low likelihood of ever being assigned a mission. As a group they compiled an excellent flying-safety record: no casualties. Did this reflect the natural caution of the scientific character? By comparison, of the fifty-four pilot astronauts selected, eight died in accidents (and only nine had not flown a mission by 1975).

Skylab was the vehicle that carried the last three members of the original scientist group into space. My interest, of course, went far beyond that of an ex-astronaut turned civilian spectator. Skylab at one time had been my baby.

Late in 1968, Deke had summoned me to his office to inform me of a new assignment. I had been mildly curious since a month before the flight of Apollo 7, when we learned that Donn Eisele would be going to back up Apollo 10 and Wally Schirra would be retiring. I was left the only member of Apollo 7 with an uncertain future.

From time to time, Deke had involved me in areas related to science, and I had a hunch what the new job might be. What I couldn't decide was whether it would be a good deal or a loser. It was obvious that Deke didn't look upon it as a plum or he wouldn't have hesitated so long in breaking the news. I suspect he didn't want to risk affecting my attitude before the flight with the news I would be inheriting the Skylab mess.

Skylab—or Orbiting Workship as we called it in those days—was the poor relation of the manned space program. It was conceived around 1965 as a way to utilize the surplus Saturn 1-B hardware (the small booster) purchased for the development phase of Apollo. The Saturn 1-B was an early, smaller, and cheaper booster which we used for the orbital tests of the hardware, launch escape systems, and heat shields. As schedules slipped and we grew bolder with each mission, some of the planned development flights were deleted, leaving extra hardware in storage.

The orbiting workshop was conceived to utilize an empty booster fuel tank as a rudimentary space station, within which crew members would perform scientific experiments. It was targeted for launching in 1968.

In the beginning, not a single flight-experiencd astronaut was assigned to the program, and the project floundered from one false start to another. Alan Bean was the first to head the Skylab branch of the Astronaut Office and (in time) was promised command of the first mission. Al was rescued from this thankless pioneering when he was called on to fill the vacated seat on the Apollo 9 backup crew, after C. C. Williams collided with the ground at a speed of Mach 1+.

In 1967, Gordon Cooper took over Skylab on his road to nowhere. Here at last was some sorely needed experience, but the program had little direction and was the first to suffer from the perennial budget cut, and Gordo's heart really wasn't in it. When he took over, it was scheduled to fly within two years. And when he left, a year later, lift-off was still "a couple of years" away.

With most of the astronaut manpower thrown into Apollo, Skylab fell next on the shoulders of Owen Garriott, the senior scientist-astronaut. Owen was conscientious and hard-working. He had a personal interest in the results: it would be his only possible chance to fly. But his was an interval when whoever ran Skylab couldn't win; he could only hold down the score. Garriott had the dual handicaps of limited experience and lack of support in our office because everyone's attention was focused on the flying program—Apollo.

He received little guidance for two reasons. First, there was little interest in an earth-orbital program that would be manned by scientists. Second, there was the astronaut syndrome of having the ultimate, last word. This attitude could be characterized loosely as, "Let them have their fun for now; when we assign a flight crew we'll straighten it out. Hang the cost. If they don't fix it, we won't fly it." To the old pros, the scientist-astronauts were just keeping the books, so they would know where the problems were when the heavies arrived.

Meanwhile, engineers from the Program Office were having a field day in areas which were formerly our exclusive province. Earth-locked engineers, with NASA since the early days of the Mercury program, were still designing for 1-g rather than taking advantage of the zero-gravity environment. As a result, at the end of 1968, Skylab was saddled with such bizarre concepts as:

1. *Compression walking,* which presumed that man would move around the workshop by pressing his hands against a "ceiling" and his feet against a "floor" and perform a movement as close to walking as possible.

2. A *cargo transfer device* requiring that a monorail system be fabricated in orbit (Tinkertoy fashion) through each of the modules between the spacecraft cabin and the crew quarters in the workshop—a distance of 110 feet. This was to enable our puny crewmen to control the large packages that would be transferred back and forth in the vehicle—all of which were weightless, of course.

3. A *pelvic restraint system* in the ward room, designed to keep us

from "flying" all over the place. It consisted of a metal arm, extending from the wall on a universal joint, which clamped securely around one's pelvis. It would hold securely, yet allow movement at the end of a boom.

4. An honest-to-God *settee* along one wall of the same ward room (I assume for those who wished to sit out a dance).

5. A *bicycle-seated, pedal-powered carriage* mounted to a pole running the length of the workshop. Through an arrangement of gears, pulleys, and extension rods, an astro was expected to pedal himself to any spot within the twenty-one–by–seventy-foot tank, and thus solve his problems of getting around in space.

If these descriptions make Skylab sound like a can of worms, don't let me mislead you. It was! It would have been easy to feel I had been exiled to a program staffed by has-beens, never-was's, and never-would-be's attempting to fly a bunch of leftover junk. My own appraisal was much more hopeful. If one didn't have a flight, the next best thing was to have autonomy, responsibility, and authority. That was almost impossible to come by in our office, but it was available in the Skylab job. The lack of interest around the Astronaut Office insured little interference, all the authority one was able to assume, and enough privacy to get the work done.

All that, and problems big enough to satisfy the most aggressive and ambitious manager. The challenge was to get a program flying that could never have gotten off the ground as it was constituted in 1968. What it lacked was operational input and effective management.

While I did not exactly do cartwheels off the nearest curbstone at the new assignment, I wasn't discouraged either. It was an opportunity to produce, and those two years at the helm of Skylab became my real contribution to manned space flight. Two events took place that made it possible to turn things around. Within a year the program manager's job was turned over to Kenny Kleinknecht, a veteran of the Mercury, Gemini, and Apollo programs. At the same time, the Astronaut Office was forced to accept seven air force astronauts from the canceled Manned Orbiting Laboratory (MOL) program. We were off and running.

As rapidly as possible, I reassigned the scientist-astros to their areas of greatest competence and credibility. For most of them, that meant working with the experiments and experimenters. They may have lacked experience, but they were among the brightest people in the office and they were working on a program whose success would be judged on scientific return. By channeling their effort into the right areas, we could guarantee that the experiments—90 percent of the mission—would be the most useful ever carried into space.

Operational hardware decisions were quickly returned to those with a pilot background for handling. That's where the seven MOL pilots came in. The Rube Goldberg concepts were eliminated. The science fiction view of living in space was made to give way to a realistic approach based on coping with the slightly off-center world that existed up there, and achieving the possible. Management was prodded to switch from the empty fuel tank workshop on a Saturn I-B to a "dry," engineered workshop launched on a Saturn V until they were convinced.

In perhaps the most important step of all, we brought together the five widely separated manufacturers involved in Skylab and took a unified approach to all of the spacecraft systems and controls. We initiated cross-pollination of ideas instead of letting each company huddle in a corner with its own piece of hardware.

Overnight I became part of that "other program" no one in the office cared about—the one that was going to fly on a date that always seemed two years away. My beaten path, my own personal orbit, now moved from such lifeless spots as Los Angeles, New York City, and Cocoa Beach to such exciting cultural centers as Huntsville, Alabama, and St. Louis, Missouri. My first trip to Huntsville was spent in a four-hour critique of the orbiting workshop mockup before the engineers could clutter my mind with all the reasons why it had to be the way it was.

At lunch that day with Wernher von Braun, I shared my conclusions, including the ridiculous concepts already mentioned, and he carried them higher. That session stirred some juices in Houston, and the program people complained to Deke that I had gone over their heads. The important thing was that Skylab had begun to move. I found myself thriving on the variety of problems that arose.

It was hard to get anyone interested in Skylab when crews were landing on the moon every couple of months. We went about our business, and the rest of the Astronaut Office went about theirs. The fellows I had entered the program with were all involved in Apollo, soaking up the glamor, the lunar landings, the world tours, and the awards. And over in the corner there were fifteen to twenty of us working on Skylab in near-CIA secrecy.

One of the small benefits of Skylab was the opportunity for periodic contact with Wernher von Braun, one of the pioneers of rocketry and a man I had come to enjoy, admire, and respect. Von Braun was one of these people who fill a room, a large man, with great self-confidence. My first exposure to him had been in 1964 when our group of fourteen toured the Marshall Space Flight Center at Huntsville, Alabama. On that occasion von Braun held court in a huge conference room, big as a ballroom, with three rear-projection screens behind him. What impressed me most during those eight hours was that no matter how technical the question, he seldom had to ask a deputy for an answer. He not only ran the complex, but he knew what it was all about as well. All through those early months of Skylab we worked with von Braun and his people at the Marshall Space Flight Center.

I think of those early months of Skylab as good months, full and hectic, wound tight with the business of pushing ahead. At the same time there was great nervousness on the part of the scientist-astronauts, a feeling that at any time the project would be canceled and they would miss their moment in space. By late 1969 their worries were pointless. Skylab had reached that enviable point at which it would be more expensive to stop than to keep going.

Now it was my turn to wonder. I went to Tom Stafford looking for some assurances that I wasn't putting in two years of hard time only to get pushed aside and have someone else take over. "What are my prospects?" I asked. "All bullshit aside. If I'm not going to fly here I want to get back into the Apollo mainline and into another flight. I don't plan to wait around developing the hardware just so someone else can cash in."

Tom knew the signs. "If you stick it out," he reassured me, "you'll be commanding the first mission."

I reminded him that the same promise had been made to Al Bean.

The pressure was building within me to get that one more flight every pilot craves or to get on with the real business of life. It was getting impossible for my family. I was traveling more, bringing home a briefcase every night, and I couldn't even tell Lo with certainty that I was doing it for a reward that was even in sight. I was gone for days at a time, and when I came home my contribution to family life was to chew the kids out the first half-hour, to maintain a sense of discipline. It was the only consistent part of our routine.

You can tolerate almost anything as long as your wife and kids love you and put up with it, too. But Lo was having trouble understanding why I stayed with Skylab when no one could even predict a flight. About this time a few of our friends suggested that Lo run for the city council in Nassau Bay and I encouraged her. My contribution consisted of going from door to door, knocking softly and hoping no one would answer. When they invariably did, I would say, "Hello, my name is Walt Cunningham. My wife is running for the city council and we'd like to get your vote."

Out of eighteen candidates for five seats, Lo was the leading vote-getter. She sat on the council for a year and a half, and our home life smoothed out as well.

Not so the job. Pete Conrad's Apollo 12 flight was history, and there was speculation about his next destination. He could either take over the Astronaut Office, if Tom Stafford resigned to take a stab at Oklahoma politics, as some thought he would, or he could wind up as Deke Slayton's assistant.

My concern was a third possibility that no one mentioned: Pete could decide to take over Skylab. He not only was senior to me, he also had the horsepower and the moxie and was respected by everyone. I was rooting like hell for Stafford to declare his candidacy for the Senate, creating a fine job for Conrad. My encouragement fell just short of making a campaign contribution. Unfortunately, Dewey Bartlett decided to run for the nomination, and Stafford pulled out.

I could see the old skywriting on the wall, but it didn't happen overnight.

By the middle of 1970, two years after taking over Skylab, I was having frequent conversations with Pete Conrad. He was showing a growing interest in how Skylab functioned. We never really got to the short hairs. I sensed what was coming and didn't want to be an ass about it. We discussed the possibility of Pete's retiring and trying one of those big money jobs waiting for us out there like grapes from a tree. But the market was down, the economy was shot, and we both knew it wasn't the right time to bail out.

I tried to convince myself that even if Pete took over Skylab, I might still get the flight, since he had already made three and this one had been "promised." Or, for sure, I'd get the second Skylab mission.

In August, word came down that Pete had been assigned as the new director of Skylab. There was no reason to be bitter toward Pete. He was taking it because it was the best job available, and he did it in a fashion that was characteristic of the way all of the astronauts worked. Each of us had confidence in his own judgment and ability. When Pete moved in he had no qualms about setting up his own system, diving right in, and doing it his way. All of which left me feeling like a fifth wheel.

When crew assignments started flowing out of Deke's office, my apprehensions were confirmed. Pete would fly the first one and my crew was scheduled to back him up. Al Bean, Pete's old roommate, would fly the second. What now? Hold out hope for a rumored flight with the Russians four years hence? It was one more straw and contributed to a decision that was almost painless. I resigned from the program.

Skylab had been restructured and was now able to fly, but on bowing out I confess to pessimism on at least one objective. Knowing the engineering compromises accepted, the poor management it had labored under at the beginning, and the impact of lower budgets on key decisions, I doubted that all three missions could be flown as scheduled. There was doubt in my mind that the workshop would operate for the full ten months of the program. I anticipated the crews would have problems in orbit—not life-endangering but an accumulation of small, nagging ones that would add up until the spacecraft would simply not be able to function.

Skylab came to its moment of glory in May 1973, and one hour after launch it looked like a death blow had been struck at the long-struggling program. Some of the electrical system and a large portion of the thermal and meteorite protection shield had been torn away during boost. My first thought was of the agency's awful task of explaining to Congress and the nation how a $2.5 billion program failed to come off. My second was of the running battle bought (and lost) by astronaut Bob Overmyer and myself with the McDonnell-Douglas engineers on the reliability of the very hinges that failed during boost.

Those thoughts came rattling back to me when I watched the Skylab 1 crew return to earth five weeks later. Not only had they salvaged the entire program with their successful—I damn near said *heroic*—efforts at repairing the electric power and thermal protection systems, but they had performed many tasks for the first time in a space environment. When they packed up and left the workshop in orbit, we could be optimistic for a Skylab 2 mission. And Skylab 3, for the first time, looked to me like a good bet.

Joe Kerwin, our medinaut on Skylab 1, became the first doctor to make a house call in outer space, but unfortunately—or fortunately, depending on your point of view—he had nothing to treat. However, he would have come in handy on Skylab 2. That first space-station crew suffered some early discomfort but nothing that really affected their efficiency. The Skylab 2 crew—Al Bean, Owen Garriott, and Jack Lousma—were pretty sick boys for the first few days, and Garriott and Bean had to tread lightly for the first week.

They started out behind the power curve but at the end of ten days they were clicking along pretty well. As they approached the end of the mission, eight weeks later, their workload was running at 150 percent of the original flight plan, a record untouched by any other flight crew.

Everything was running so smoothly that they began a small campaign to extend the mission past the originally scheduled fifty-eight days. They would call home twice a week on a VHF radiophone connection to convey impressions and keep up with family activities. Some of the conversations were reduced to standard husband-wife dialogue. On the crew's forty-second day in orbit Sue Bean, Al's wife, called to invite Lo and me to a going-away party for the Dave Scotts. I asked Sue how it was going.

"Oh," she said, "I'm in the middle of cleaning the darn swimming pool, the lawn is out of control, and Amy has a cold. It's sure getting old, and today they asked for a ten-day extension. I sure hope they don't get it."

I had to laugh. Al was away on "man's greatest adventure" and Sue was complaining about cleaning the pool.

The third Skylab spacecraft carried an all-rookie crew, the first since Gemini 8. Gerry Carr, Bill Pogue, and Ed Gibson had their hands full. The structure and logistics of crew training routine is tremendous in scope. It takes place at sites spread all over the country, sometimes throughout the world, and calls for a great deal of discretionary judgment by the crew commander. Pete Conrad, as the chief of Skylab, faulted himself for not spending more time after his own mission working with the Skylab 3 crew on training priorities.

Whatever the failing, the fellows had their troubles right from the start of their three-month stay. They didn't feel well, suffering a rather well-publicized siege of first-day barfing, and ran continually behind the flight plan *and* then behind the revised flight plan with its reduced workloads.

Ground controllers and the public became aware of the loaded barf bag in the same fashion as on Apollo 8—the on-board tape recorder. In that earlier case, it was by design; on Skylab 3 it was an unfortunate accident. In all live communications they had chosen not to mention that Bill Pogue was sick enough to throw up. Their mistake was in discussing what to do with the bag and what not to tell the ground while the on-board tape recorder was taking down every word. The tape was later routinely "dumped" to a ground site and many hours later replayed at Mission Control. Flight directors, medical doctors, and management all the way up to Chris Kraft were understandably less concerned about the vomiting incident than they were about the subterfuge. The riot act was read to the crew, but the incident compromised both their credibility and their relationship with the ground for the remainder of the eighty-four days.

It was a tough rap for the crew because it was not a self-serving decision, even though it appeared that way to observers. The period just before their flight, in late 1973, was marked by congressional budget hearings on funding for the space shuttle program. NASA was happily describing a twenty-year program of eighty flights a year, each lasting from seven to thirty days. Some legislative critics were just as blithely knocking the program in the head, especially the planned seven-day missions. "Hell!" they said, "it takes a week before the crews adjust enough to perform useful work. On the second Skylab mission it was nearly ten days."

All this was in the papers. NASA could not refute some of the criticism, but they were placing great emphasis on medical means of adapting to zero g as early as possible. It was a very sensitive issue. No one at NASA told them to be deceptive, but the fact remains that Carr, Pogue, and Gibson believed they were helping the agency in an hour of need. It was even discussed on the same tape that let the cat out of the bag on the vomiting incident. That part of the accidental communication was either suppressed or not considered newsworthy, and it was assumed that the crew was merely protecting a macho image.

Carr's crew recovered from their jumpy stomachs, and what had originally been planned as a fifty-six-day mission was stretched to an important eighty-four days. The endurance mark set by Pete Conrad in four missions was surpassed by Alan Bean in two, then placed out of reach for many years to come by three rookies on their first, last, and only mission.

Through knowledge gained during the earlier missions and adjustments made to the on-board regimen, the third Skylab crew was in much better physical shape on return than the second. Their accomplishments give us greater confidence that when the hardware is available for travel to the planets, man will not be found lacking.

Apollo, the moon flights, and all the steps leading up to them were for the poets, for the ages—a climax to the forty centuries in which man had dreamed about and finally reached the moon. But Skylab was for people; it was for *now*. It was what the champions of space exploration had promised would be there when the United States ended its celestial trade with the moon—our billions for its rocks.

Much of Skylab's success was the result of contributions made by the scientist-astronauts. Their disciplines were more applicable to Skylab than Apollo, but more important, Skylab was better structured operationally to accept their contributions. Their own reactions were mixed.

Don Holmquist, the M.D. who was an early dropout: "The most stimulating group of people ever, but overall, it was a stifling experience." He resented the way we responded to the space-flight carrot dangled before our noses. "Top management," he added, "was incompetent."

Phil Chapman, a member of the infamous XS-11, was a scientist in spirit as well as in fact. His career moved from MIT to NASA to a research position in industry. He was among the first to take a realistic view of the scientist's lot in the Astronaut Office, yet he stuck with it for five years. "At NASA," he said, "I was out of the mainstream of science, but I don't regret the time."

When I suggested that the National Academy of Sciences had been premature with its political pressure, Phil disagreed. "They should have proceeded with the selection on an assumption that the lunar landings would be easy, meaning scientists could fly sooner. The first group should have been solely geologists, for the lunar missions. NASA was a classic example of the system: management with total control and the astronauts with no bargaining power."

I agreed that the first scientists should have been geologists, but I'll be damned if a lunar landing was ever easy. It was always the *first one* for the crew that was making it. Astronauts actually had a great deal of power; it was in program management that we had little responsibility or authority.

Owen Garriott's first mission, Skylab 2, came after an eight-year wait. "The whole thing should have taken half that long," he said. He felt, on balance, that the eight years were well-spent, but he was not completely satisfied with what had occurred in that time. He thought the flight crew training had been overdone—but after eight years, who wouldn't?

Owen went on, "None of us really change after a flight—we only have the freedom to do more of what we want. Before the mission, each one has to concentrate so much on a single goal he is unable to be an individual."

In spite of the successes of Skylab, the public remained completely apathetic. Part of the problem from the outset had been the inability of the masses to understand exactly what was being done and what could be achieved. Explanation isn't easy when most people are still trying to figure out how a radio works. But even the average taxpayer can—or should—understand the need to observe our lands and oceans and to control our environment.

The casual attitude toward Skylab could be interpreted as confidence in NASA, but the course of the program was anything but routine. It may well have been NASA's finest hour, challenged only by the successful return of the Apollo 13 spacecraft. Had we been unable to send crews into Skylab, we would be designing our next generation of space stations (and selecting crews) without ever subjecting man to a really long stay in space. Man's investigation of the sun and of the earth's resources would have been delayed at least another seven years.

The phenomenal capability of NASA's resident experts was shown during the thermal shield failure of Skylab 1 when they correctly diagnosed the problems from 270 miles below. In less than ten days, apparatus was built to solve the problems plaguing the spacecraft, making it habitable and useful for the ensuing ten months. Conrad's crew was launched in time carrying with them one primary "fix" and two backup solutions.

For the astronauts, we saw it as a revolution of sorts in the thinking of planners and flight directors. Historically, flight crews trained long and competitively on all phases of a mission and were not allowed to perform any task which had not been through a lengthy certification and practice cycle. We were overtrained to the point where even crucial operations seemed routine. Now, overnight, it was accepted that Conrad, Kerwin, and Weitz would perform all manner of circus stunts under conditions formerly believed too risky even to consider.

The successful repair of Skylab 1 emphasized the value of real-time decision-making by broadly trained crew members. It was an effort that was incompatible with the Russian philosophy of manned space flight.

Dollar for dollar and pound for pound, Skylab was one of this country's most important space achievements: for what it added to present knowledge in solar astronomy and material processing, for the new ground it broke for future exploration of the earth's resources, and perhaps most important, for the study of man's long-term adjustment to the environment of space. I look back on it as a great comeback from the edge of disaster and one of the space program's brightest moments.

15

The Dispensable Astronaut

SYSTEMS ENGINEERING has been called the science of trade-offs. By the early 1980s we will be flying a few women in space, and when we do those first few astronettes will be the result of trade-offs.

There is just *no way* to get into this subject without sounding like a grand champion male chauvinist, yet I don't believe God formed women out of a pound of barleycorn. Nor do I wish they would all quit being hysterical and go back to decent jobs—like working as manicurists.

There is no doubt in my mind that they can do many of the tasks in space as well as any man—*or* that it will cost more, create new risks, and further complicate the operation. Logically, when the added costs and other mission impacts are balanced by the benefits, then the proper time will have arrived for a woman to take her seat in the capsule.

Logic, of course, will have little to do with it. We live today in a society ruled by lobbies. Women will fly in space long before the time is right simply because social pressures will compel it. There will, in fact, be only token opposition. The public will consider it fair, the politicians will say it is the law, and those most intimately involved, the other astronauts, will find the idea of a mixed crew . . . well . . . intriguing.

Voyages to other planets in our solar system will be measured in years, not days, and one is more likely to attract volunteers—among the astros I know—if the crew is coed. Of course, mission planners will expect all members of such long-duration flights to be qualified in special areas of science. But honesty compels me to point out that most of the eminent women scientists are not renowned for their *Playboy* centerfold bodies or their sophisticated views on sex. So here we have the beginnings of a conflict basic to the human experience. After, say, a year in space, your average male astronaut could fall in love with an avocado. But the idea of being used or even thought of, in romantic ways is frequently unacceptable to the liberated woman's view of equality.

This isn't exactly the major problem confronting the space program in the next ten years, but he-she chemistry being what it is it can't be ignored. A spaceship is a damned small room and, even under the most detached circumstances, you get to know each other better than you want.

Long ago I lost count of the number of times one or more of us has been cornered by flag-waving feminists: the woman physicist who had competed with men for years, the pilot who felt that her flying skill was enough of a qualification, the physical culturist who was prepared to arm wrestle, winner take all. Each had the same challenge: Why are there no women in space?

My reaction is always the same: Are they motivated by a desire to challenge the unknown . . . or simply to challenge what they see as the male establishment? They frequently argue that, "The Russians sent up a woman, why can't we?" But the question isn't why can't we or why didn't we, but why *should* we? The Russians pride themselves on being able to launch, and recover, their spacecraft through completely automatic methods. This has enabled them to fly dogs, civilians, and even a woman parachutist named Valentina Tereshkova.

The American space program doesn't operate that way. We have allowed, and the astronauts have fought for, a higher degree of human involvement and control of the spacecraft. The thought of an automatic deorbit and recovery is no more appealing to an astronaut than the idea of an automatic landing is to a passenger on a commercial airliner. Any time the job at hand can be best performed by a woman and, other qualifications being equal, a man gets the assignment, I'll rise up on my hind legs and shout, "Foul!" I'll be on her side. But if a woman is launched in space merely to demonstrate that we can do it, then I will denounce that as the most blatant form of tokenism and a poor justification for spending the taxpayer's money.

To those of us who bought the tickets and took the rides in the first ten years of the manned space program, it never occurred that it had to be *socially* relevant.

It was inevitable that the character of the original space pilot corps

—the seventy-three crewmen selected between 1959 and 1969—would in itself become an issue. It was hard to ignore what they had so visibly in common: they were all white, Anglo-Saxons. Except for a few Catholics, you could describe them as Wasps. They were nearly all middle class and mostly from small towns.

I don't know why there haven't been any black astronauts. For that matter, I don't know why there haven't been any Jewish astronauts, or Chinese or Latins. I only know that the absence of a black astronaut was not due to prejudice *or* to a lack of pressure from the highest political levels. In the early and middle 1960s, it was the fervent hope of those in the space agency that a well-qualified black would apply. That might have solved the political problem painlessly. But to be told to "find one" and, if necessary, to stretch a point to enroll him, was an affront to the many extremely able men who had already been eliminated. Beyond that, it would have been repugnant to the men entrusted with the selection process. I can only guess at Deke Slayton's reaction at being told to bend the requirements in order to desegregate space. "What do you mean, I *have* to take a black astronaut? We take the best we can get, and if one qualifies we'll be glad to have him—if not, and they still insist, they'll have to find someone else for my job." Deke's only prejudice is against failure, but he is one of those people who resist pressure. If someone insisted on lowering the standards he would tell them in a most clinical way how far to lower them and where, and then he would advise them to find another man to do the selecting.

These are not statements casually made. We went through a period in the late sixties when even the word *qualified* became a kind of code word for racial discrimination. On that score I don't believe the space program really needs defending. As late as World War II there were only a few black military pilots, and it wasn't until the late 1950s that a major commercial U.S. airline began hiring black pilots. That was one of the factors working against them when the United States was ready to explore other worlds. For a long while, the men who flew, who wanted to test airplanes, and who were motivated toward space flight were white.

Aviation simply didn't attract minority applicants in any real numbers until the war in Vietnam. During my tenure in navy flight training, in 1952 and 1953, only three of my contemporaries were black. Out of 300 qualified applicants for the fourteen astronaut openings in 1963, there were only two black military jet pilots.

To my knowledge, an able young air force captain named Ed Dwight came closest to breaking the barrier. In the early 1960s, Dwight attracted attention not only in the air force, but also within several groups actively working to break down racial barriers in the United States.

When Captain Dwight was attending the Test Pilot's School at Edwards Air Force Base, not very subtle political moves were taking place to see that he would also attend the Aerospace Research Pilot School and, in the process, be duly qualified to become America's first black astronaut.

Fifteen years before all this, another young air force captain, Chuck Yeager, became the first man to exceed the speed of sound in what was then the farthest reaches of manned space. Now, as a colonel, commanding the Aerospace Research Pilot School, his career had crossed paths with this appealing young captain who hoped to penetrate the boundaries of space.

I don't know how or by whom, but it was proposed to Yeager that a special effort should be made to include Dwight in the Research Pilot class then being formed. Yeager's response—as I got the story—was in character with the man and his sense of fair play. "We will take Captain Dwight," he said, "but if we have to lower the standards to do so, I will also enroll all those between the usual cutoff point and Dwight's class standing." Not coincidentally, the next class at the Aerospace Research Pilot School—usually eight in number—was the largest in history with thirteen members. And not too long after that Dwight was way down among the also-rans in the selections for our group of fourteen, the Apollo astronauts.

I'm sorry Dwight didn't make it. From what I was told by people who knew and observed him, he was a capable aviator. His appointment would have pleased those who saw it as another battle to correct society's wrongs; it would no doubt have been splendid public relations.

Four years later Bob Lawrence, an air force major, was selected into the Air Force Space Program. The first and only black astronaut was killed in the crash of his F-104 only a few months later.

In spite of calls to NASA from then Attorney General Robert F. Kennedy, Dwight was not selected, and for honest reasons. Every improvement, every success, every new horizon in the history of flight and space flight has been governed by one rule: for a technical job, you get the best technician available. The whole idea of tokenism, or a quota system, is alien to a program that sets out to challenge the best that man can muster. Keep in mind that of all the qualified applicants for the astronaut corps, 95 percent were rejected. To my knowledge, the selection process has always involved measuring the applicants against each other and against the requirements of the job in complete disregard of an individual's race, color, creed, or national origin. It has been, and I hope always will be, damn difficult to be chosen whether the candidate be black or white, male or female.

One point needs to be made here. All during my early years in NASA, I *never* heard the subject of race discussed as a factor in the astronaut selections. It was not an issue until the politicians raised it.

It is, on the face of it, a contrived issue on the order of demanding that women and blacks be allowed to compete in the Indianapolis 500. They have the same right to risk their necks as anyone, so long as they can keep up with the field.

Of course, it isn't possible to talk about such things without sounding as though someone were trying to integrate the Masonic Lodge or liberate some exclusive, all-male bar. When minority pressure groups zero in on the space program, they invariably make it appear that someone is being deprived of a basic human right. The public always pictures an astronaut in terms of the flight itself and the attendant glamor. Few stop to consider the four to eight years' grind necessary to get there. It takes that long because of the nature of space exploration. Our decision to prepare for the flights in such a meticulous fashion has proven to be correct. There will be some changes and the time will be shortened, but it will continue to be a marathon of preparation. The selection process, as much as anything, determines the person with the drive and dedication to endure those demands.

The long training ordeal is not without hazard. No astronaut has been killed in space but we have lost eight during training. Perhaps we were spared a fatal accident in space because Grissom, White, and Chaffee paid the ultimate price in the fire on Pad 34. The fact that they could, and did, accept the risk was essential to the job. In this respect, being an astronaut is as much a state of mind as an occupation. The fire on Pad 34 and its aftermath became a part of our cargo, one of those flash happenings that shape us in ways we cannot always explain. But it reinforced my feelings that there has been no place for women and their view of life in the space effort *thus far*.

Grissom, White, and Chaffee were each survived by a wife and several children, and all accepted their loss according to the code—for a while. A couple of years passed before Betty Grissom filed her lawsuit against North American Rockwell, its subcontractors, and the space agency over the death of her husband. Betty had her reasons and we—Gus' teammates—were in a poor position to judge how imperative they must have seemed to her. But around the office the reaction was nearly unanimous: Gus Grissom had to be spinning in his grave.

Perhaps our view of it was simplistic. The country hadn't recruited us to take the risk and accept the rewards *if* successful—and file a lawsuit if it was not. We were professional pilots whose entire careers were based on assessing the risks, minimizing them, and tolerating what we can't control with open eyes. No one was more aware of the poor state of the original Apollo hardware than Gus Grissom. The astronauts' failure to stop the schedule and demand that the flaws be corrected was our biggest collective mistake during my eight years with NASA. But our own attitude—the fighter pilot mystique, "We can fly the crates they came in"—kept us moving toward a first launch date that could not be met except with disaster.

With all that, Gus would have been the last to ask for retribution. Betty saw it differently. She sued. And we claimed it was a breach of faith with Gus. At least that's how I felt about it. As a realist, I had held my breath for several years after the fire, expecting someone to sue somebody. We live in a suing society and that was an ideal situation given the known condition of the spacecraft, the conclusions of the investigation, the political climate, and the concerns of a large company for its public image. North American Rockwell settled the suit out of court in 1971, for $350,000. It happened rather quietly, a few paragraphs in the paper without the public noticeably recoiling at the idea of a cash value being placed on an astronaut's life. A short time later, at the initiative of North American Rockwell, a similar settlement was bestowed on the widows of Chaffee and White.

All things considered, Rockwell was undoubtedly relieved to close the books on a tragic episode for such a nominal amount. The company never tried to defend itself against many of the charges made against it—some totally wild and unfounded. The settlements to the surviving families came long after the spacecraft had been transformed into one of the most remarkable pieces of machinery ever built by man.

In a way, that action, that lawsuit, the idea of a hero's widow going public contributed to the transformation of the astronaut image then taking place. To be sure, other factors affected it more, including the fact that NASA itself was in a state of flux and undergoing a period of self-criticism—and, of course, the Apollo 15 envelope caper had shaken the tree. In earlier days, the astronaut image had been rather easy to control. But there were too many of us by 1970, too many sources for leaks and cracks.

As difficult as it was to get into space and join those exclusive ranks, getting out wasn't all that painless, either. Even so, by 1970 the exodus of older astronauts was nearly equal to the rate at which we were recruited eight years earlier.

It wasn't until months after leaving the program that I tried to define the reasons for my own resignation. It was something one *had* to think about: the subject came up in every interview. It boiled down to *family*—looking up every few months and realizing that my kids had grown a little older without me—and *pride*—not having the prospect of another flight, not feeling useful once Pete Conrad took over the Skylab group.

Deciding to leave was the first step. Once that decision was made, I gave plenty of thought to how I could make use of all those years as an astronaut. I mean that in the sense of taking advantage of one's education and his other experiences. Of course, to best enjoy the fallout one needed to follow the Bill Anders Principle: be the only astronaut in town. As Anders once put it, "If the governor wants to invite an astronaut to the ball, you don't want to be in a situation where his press secretary says, 'Which astronaut?' "

Even in putting that life behind, it was plain that the competition never really ceases. Just as we kept a jealous eye on the crew rotations, we followed each other's fortunes in private life: books, jobs, business ventures, second marriages. It was the kind of checking one does between class reunions. We would not become old men, living on past victories and regretting old defeats, even though some are still fresh in my mind. The years at NASA of watching a new crew come onto the scene, move to center stage, and then fly in a few months sometimes hurt. There went another mission you'd never fly. As someone once said, "Every time a friend succeeds, I die a little."

As each of us faced the dismal prospect of no more flights, or when we concluded that the prize was no longer worth the game, we were faced with several options. One might retire in place, staying in the Astronaut Office, retaining the astronaut label but performing little of what had earned it. It was a shell, a sham, akin to the "partynaut" we often joked about. It would mean attending a design meeting now and then or making a speech. Your astronaut career, as a practical matter, would simply have died without benefit of burial.

Some of the astros tolerated such an existence because they had reservations about making it on the outside. Many of the military astros used it as a holding pattern to complete their twenty years before retirement. Others chose that role because they liked it and were content to soak up the suds of their celebrity status, like the old boxing champ who allows the hangers-on at the bar to buy him drinks.

Still others chose to remain in the public sector, either in another capacity at NASA or with a different agency of the government. The most successful at pursuing this path were Jim McDivitt, as the Apollo program manager for most of the lunar landings, and Bill Anders, who went on to become the first administrator of the Nuclear Regulatory Agency and United States ambassador to Norway. Mike Collins was named director of the National Air and Space Museum, and Dave Scott became director of the NASA Flight Research Center. Others—

such as Al Worden, Gene Cernan, Rusty Schweickart, Jack Schmitt, and Jack Swigert—found desirable government jobs with status and still within the smell of the greasepaint.

A third way was to move into the private sector of business. This was sometimes determined by office politics or the lack of an available government job, but for most it was the challenge of an entirely new career and the desire to make money. When we went out to sell our services it was, as in any commercial enterprise, a case of *caveat emptor*.

Far too often, unfortunately, the potential employer saw us as neon signs that said A-S-T-R-O-N-A-U-T in flashing lights, but not as individuals. That group image carried with it a blanket assumption of certain skills that was not always warranted. Not only did our abilities and technical knowledge vary greatly, but in many cases so did the willingness to dig in and work, to slug it out in the trenches of management with such details as monthly reports and personnel problems. In some instances there was a plainly detectable attitude of "I went to the moon, and now the world owes me a living."

It is reasonable to assume that most of the astronauts are well equipped technically, that we have a certain credibility with the public, and that the name alone will open many doors. If we were all run through a computer—which many seemed to think we were—the printout would probably describe us as goal-oriented, quick-study types, with a high need for achievement. Most have strong leadership qualities, although many lack specific management skills. Most of us fit somewhere between a professional manager and an entrepreneur, with a reliance on technology but even more on ourselves and our ability to solve problems.

None of this was yet worked out in my mind that day, in March 1971, when I stopped by Deke Slayton's office. After considering my resignation for several months, Lo and I had finally made our decision. I was leaving the program and asked to be relieved of my administrative and technical responsibilities. It would let someone else gain experience by picking up my projects and leave me freer to investigate the job market. I had already done some investigating around the country, and it would only have been a matter of weeks before Deke heard it from someone else. Deke thanked me for letting him know and indicated it was a little unusual to be alerted in advance that someone was quitting the program, especially if they had yet to decide where they were headed.

That was within a few days of my thirty-ninth birthday, and I did not agonize over it. I nodded and spoke the words to Lo, "Honey, we're through playing space," and she gave me a smile like sunshine flooding a window. From that moment on it was only a question of when and where.

When asked, "What's it like to be an astronaut?" I frequently answer somewhat facetiously, "Everybody's got to do something for a living." I made a conscious effort to be as little changed by the experience as possible. It's impossible to be completely unaffected, but I like to think that my basic characteristics remain unchanged.

I was exposed to things I would never have experienced otherwise. And it happened at the right time in my career. Retiring three years later, the world outside NASA would not have offered the same opportunities to Walter Cunningham. That was part of the consideration in deciding *when* to leave. Being an astronaut is a good job, but it's not the living end. It took five years to reach a position (in the Skylab program) where I could materially influence the destiny of a multi-billion-dollar effort.

I came to Houston with few preconceptions about the astronauts. It was encouraging to find out they were no better or no worse than anyone else, full of the same human foibles, and that my credentials were not out of place. We had ass-draggers in the program along with workaholics like Slayton, McDivitt, and Scott.

The economy wasn't exactly roaring along at peak efficiency in 1971, but that concerned me little. It had never been my plan to fly rockets the rest of my life. The decision to leave fit the general plan that had taken shape years before: to spend my twenties in education, my thirties in space flight, my forties in the commercial world making a secure future for my family, and in my fifties to consider public service once more. I hoped that by then I could afford that luxury, and with the children grown I would feel less guilty about asking my wife to repeat the trials of living with someone whose life wouldn't be competely his own.

I talked with companies in insurance, oil, and microelectronics, and even considered opening my own consulting firm. There was no shortage of opportunities, but many lacked substance, being of the marketing or door-opening variety. It was the operating end of a business that interested me. Sure, it would have been an advantage to land in a high-technology spot, but as matters developed, I went into an entirely new field right there in Houston.

The space corps had been a tremendous finishing school. I joined it with an education in science (an essential qualification) and elected to use that training as a tool in a space career. After eight years with NASA there were new tools and a new confidence to use them.

No career lasts forever, but it can provide new tools to be used in the next. Being a part of "man's greatest adventure" was but one of many stimulating challenges, and it could have passed me by through any number of circumstances. Sure, a certain amount of good fortune—luck —enters into it. But without the right preparation one can never take advantage of good fortune. Luck occurs a lot more frequently to the man who is prepared.

Once my connections to NASA were severed, it was complete. Except for an occasional call to an old friend, like Stafford or Bean, my view has been from the grandstand. Long ago it became apparent that most people assume a former astronaut continues to act as a kind of unpaid consultant, that his vast knowledge is available to the government when needed, that the tremendous investment in his training and education is just waiting to be tapped, again and again. To my knowledge, no former astronaut has ever been called upon by NASA to assist it in some hour of need or even to express an opinion about any internal matter once he had left the fold—certainly not in that small corner of the NASA universe called the Astronaut Office.

It will be interesting ten or fifteen years from now to look back at what happened to the men who were brought together to form the corps of space pilots, those who bridged the gap from the beginning of our nation's space effort to the Russian linkup of 1975. During that period the astronauts were king of the mountain. They were heroes without blemish or equal, a condition that will not exist again. When historians make their judgments, the glory of the space program beginning will be found among those first thirty and, to an even greater extent, the first sixteen.

By 1963, when my group was selected, the character of the candidates was beginning to change. In order to insure an adequate number of motivated and qualified applicants, and still remain highly selective, the test-flying requirement was replaced by a flight-time experience that could be met by jet fighter pilots. As a result, half of the fourteen Apollo selectees were not test pilots, myself among them. Although our group had logged more jet time, on the average, than the original seven astronauts, it was about half the experience of the second generation, the men of Gemini. There has never been much doubt in my mind that those nine—the second group—were the best all-around men chosen for the job. That was in 1962, and Young, McDivitt, White, Conrad, Stafford, Borman, Lovell, and Armstrong (the last in his group to fly) were the heart of the Gemini project and filled most of the key slots for Apollo. (The ninth crewmate, Elliott See, was killed before he could fly in space.) It just might be that eight people more capable of performing a difficult task well, under pressure, had never before been assembled.

Time served to wipe out most distinctions between the groups, but in the beginning the Original Seven were just what the times and the stakes demanded. They were highly competitive and attracted by the glamor and renown of being involved in a first-of-a-kind venture. In 1959, the man-in-space program needed pilots who recognized the mission for what it was and who understood the challenge. At the time many better-known test pilots wanted no part of the action.

Of the first three groups of astronauts, twenty-three lived to fly in space, and between them they were involved in all the firsts the manned space program has experienced. In spite of this the most dedicated had to be the last three of the six groups—the last of seventy-three astros chosen from 4,000 applicants—for whom the rewards were largely uncertain and the glory pretty much used up. The public was beginning to grow bored with our space achievements. White House visits, world tours, and personal story contracts were no longer among the fringe benefits. Scientist-astronauts like Joe Kerwin and Ed Gibson trained for eight years before Skylab offered them a chance to fly. The siren call of adventure had to be loud and clear for them.

From Buzz Aldrin's nervous breakdown and Ed Mitchell's experiments in ESP to John Glenn's election to the Senate, the public's interest in the astronauts continued beyond their tour of duty. As some floundered in business, it became evident they were not much different from anyone else. Without the stimulus of space and the challenge of other horizons, some lost the self-discipline that accounted for their success as astronauts.

It is, I suppose, much like being a college football hero. But you can't go on replaying old games forever. As for myself, I left without regrets, secure that the time was right to move on, to find another direction. One year I had been away from home 255 days and my family was in danger of becoming a matriarchy. Some of the slick magazines depicted the astronauts as the new sex symbols, but I doubt that many of the wives thought of their husbands as sexual athletes. More often than not he came home bushed, carrying his work and his laundry, and probably needing to stop by the office on Saturday and Sunday to straighten out the mess that had collected while he was gone. Our children grew up in the space program and took it for granted, but suddenly they were eight and ten years old and celebrity fathers wondered if it was already too late to reclaim them. Finally, we ask ourselves, "Was it worth it?"

I only asked that question once, the day I made up my mind to leave, and the answer was yes. I achieved a goal that had consumed my life for eight years. The square was filled, I had compromised with myself less than most, generally said what was on my mind, and yet still felt constrained by the organizational straight jacket and maintaining the image. I can only imagine how those who conformed must have felt. Even a free spirit like Scott Carpenter once had to appear at a supermarket opening in St. Louis to accommodate a Missouri senator.

While actively training, there was little desire and even less time to speak out on controversial subjects. We were so isolated in our own careers and consumed by our own goals, that reading about the great events that rearranged the country in the 1960s was like reading about the problems of some Latin American government. But for the *ex*-astronaut, no good reasons existed to remain quiet. To the contrary, there was motivation to speak out, especially for those who could use the exposure (to run for office, raise money for a cause—or publicize a book).

Leaving in the spring of 1971, I had the feeling I was beating a rush. There was another factor. With the cards falling the way they were, the only way I could get another space flight would be through some calamity to a prime crewman. After eight years, the thought of waiting around like a vulture for something terrible to happen to a friend wasn't very appealing.

So the break was made. Yet in a very real sense it still goes on. It's as if we had belonged to some kind of church, and though we no longer attend services, we are still considered part of the flock. In one way or another, we continue to exploit the fact that we were astronauts and to share in the benefits in one way or another. Sometimes it is for personal gain, which isn't necessarily wrong, and sometimes for personal causes, charities, and so on, in a way that is both tasteful and altruistic.

Schweickart involved himself in a program in Houston called Crisis Hotline, an agency that encouraged calls from troubled youngsters and tries to help them with their problems. He would have manned the phones and raised money for them no matter who or what he was, but being Rusty Schweickart, astronaut, didn't hurt. The two of us helped start an environmental concern organization, and many a local and national charity drive has been headed by an astronaut. Their contribution is almost invariably their time and name, because astros aren't rich in spite of what the public may believe.

In an age when movie stars and athletes can sell cologne and electric razors and floor wax and panty hose, it may have been inevitable that many of the fellows would cut a television commercial or two. Ours is a cynical era. Remember the immortal words of Lenny Bruce: "It's time to grow up and sell out." The first commercials were by Scott Carpenter for Chevron gasoline, and Wally Schirra for the railroads. They were followed by several others—Jim Lovell and Buzz Aldrin, and even Neil Armstrong for General Time. Each in his turn was criticized by those still in the Astronaut Office, purist and hypocrite alike. Some of those critics have even gone on to commercial efforts of their own: Pete Conrad, for example, for American Express.

It isn't difficult to figure out why. It's for the money, ladies and gentlemen, for the money. Astronauts, as a rule, don't get rich, but they do become accustomed to a relatively high standard of living. All commercials exploit the notoriety of the personality making it, and in the interest of full disclosure, let me say that I did several spots for the Houston Police Department and a commercial extolling the virtues of Houston. Astronauts know this as well as anyone, but we aren't without scruples, or standards of taste. It was years before anyone picked up the rumored $100,000 for making a Tang commercial, and several of us turned down a potential $150,000 deal to advertise a cold remedy. Then there were the space-age, charcoal-impregnated, odor-absorbing socks. For most people there is a price, but each of us is concerned about what our peers may think.

Buzz Aldrin was concerned enough, before he agreed to do his infamous Volkswagen commercial, to ask the deputy administrator of NASA, George Low, for an opinion. I understand that George answered him very simply, with a long, slow look, "Well, Buzz, it depends on who you want to be like, Jimmy Doolittle or Wally Schirra."

Of course, there are no indispensable men in the space program. One by one we leave, and the flights and the projects go on just as efficiently. For all it matters, we could be living on the moon. But for those of us who actually stood there, or were in the thick of things, the lunar dust is on our sleeves and will always affect our perspective.

I am frequently asked what is the most apparent difference between my experience at NASA and what I see today in industry. That's easy. In the outside, nine-to-five world of business, people seldom take the same pride in their work—it's no doubt easier to get inspired over sending someone to the moon than over the daily order of widgets. Within NASA in the 1960s everyone took pride in their job, you could count on them—from the top managers to the guy who fastened the last snap on our flight suits.

Pride and teamwork were more than words, and that's the message I took with me. It will still be there after the mistakes and phony sentiments of an earlier time are forgotten, when the scientists have phased out the hotshot pilots and women have ridden the fire into space.

16

Giving Away Victory

IN TEXAS, where real estate deals and oil syndications are as popular as fried pies, they are often structured as "joint ventures." A joint venture, according to one cynical definition, is a partnership in which one party has the money and the other has the experience. When the deal is completed, their positions are exactly reversed.

Which brings us to the 1975 joint venture between the United States and the Soviet Union—that celebrated "handshake in space." Let me say this as fairly and gently and diplomatically as I can: in terms of who gained what from the Apollo-Soyuz mission, we were had!

The flight was an artistic success, even a political and show business success. Don't knock that. In view of what else we could be doing to each other, those are not bad successes. Photographs of the two crews working together looked terrific on front pages around the world, and it must have sounded, well, different, when Tom Stafford spoke Russian with an Oklahoma accent. That alone was nearly worth the price of the trip.

It was only in the "secondary" objectives, such as flying the spacecraft, and in certain technical and scientific goals, that we may have failed. In theory, both sides gave and benefited equally from this space odyssey that became the very symbol of detente. (Funny about that. Six months after the Apollo-Soyuz Test Project [ASTP], our foreign policy with Russia had grown so cloudy that detente had become a nonword.)

What exactly did either of us get for the nearly $150 billion the two world powers have spent to penetrate this new horizon? Was there really anything we stood to learn from the Russians? And would they have let us if there was?

Although the two programs were similar in many respects during the sixties, theirs got off the pad earlier because of the big boosters their military had confiscated from the Germans. Like us, the Russians have flown three generations of manned spacecraft. They began with the Vostok, with the late Yuri Gagarin flying the first of the six orbital missions (and Gherman Titov the second) nearly a month before Alan Shepard rode his Mercury 7 into space for a fifteen-minute suborbital flight. It was nearly a year later before John Glenn circled the globe three times. Yet in this country we frequently refer to Glenn as "the first man to orbit the earth."

For years the Russians rang up firsts like nobody's business: the first satellite, the first man in orbit, the first woman in orbit, the first two-manned spacecraft, the first three-manned spacecraft, and the first extravehicular activity. They got our attention and were politically significant even if they were sometimes lacking in technical achievement. We tried to rationalize those performances by telling ourselves that our greater concern for human life caused us to move more deliberately.

So consider this. When Glenn flew his first manned orbital mission, our experience consisted of three orbits on the Mercury spacecraft, two of them by monkeys and the other unmanned and automatic. Before Gagarin's flight, the Russians had logged approximately one hundred orbits in their Vostok spacecraft, including several missions with dogs.

They, too, had their catastrophes accompanied by interruptions in the program. In early 1967, when Vladimir Komorov, flying Soyuz 1, died during reentry due to a parachute problem, it was a year and a half before their next flight. Their program was subjected to the same painful reappraisal and redesign ours went through after the fire. Finally, after one unmanned test flight, Soyuz 3, manned by General Georgi Bergevoi, lifted off just two weeks after the ascent of our own Apollo 7. Another mission in the Soyuz series suffered misfortune in July 1971. Soyuz 11 lost pressurization when the crew separated the descent module from the oribital spacecraft. It was a crushing blow! After a record twenty-three days in orbit the entire crew of three was killed on reentry. Their manned program was interrupted for more than two years.

The Russians always seemed to encounter their greatest difficulties when returning their vehicles to earth. For fifteen years, stories have persisted of cosmonauts landing in the snow-covered wilds and not being recovered for more than twenty-four hours. These are believable

—and not too surprising—when you consider they have no on-board computer and only an extremely limited maneuvering capability to hit a designated touchdown site.

Until 1976 the Russians recovered every manned spacecraft on land, a situation ideally suited to their unique requirements. Among them: basic hardware design, a great land mass within their national boundaries, and the Soviet penchant for private, unobserved operations. Aborts during launch from a secret base in the middle of their country required a land landing, so why not on normal reentry? We applied similar logic to U.S. launches over the Atlantic Ocean. Any Russian lunar landings will require them to develop a water recovery technique in order to live within other operational constraints.

Why doesn't the United States make land landings? *Any* spacecraft landing system must be thoroughly tested prior to its actual use. It must be able (with repeated success) to touch down on the hardest surface in the recovery zone and still protect the crew. Since water accounts for the majority of the earth's surface, the craft should be capable of withstanding an emergency landing in the ocean. And since testing in the sea lends itself to repetition better than a land surface, it was an easy choice to design around a water landing. The Apollo spacecraft is capable of safely flying to and landing in a plowed field roughly the size of a large municipal airport—but what if it missed?

How have the Russians overcome land-landing problems?

During the early missions in the Vostok and Voskhod spacecraft, the crew members would abandon their vehicle around 10,000 feet and resort to a personal parachute for landing. The Soyuz spacecraft, however, incorporates a landing rocket system that reduces the rate of descent when close to the ground to a safe level. This eliminated the need to carry personal parachutes. That landing rocket is fired at eight to ten feet above the ground, slowing the vehicle for a soft touchdown. The Apollo spacecraft utilizes three huge parachutes for its water landing. I am neither the first nor the only astronaut to have reservations about a landing rocket system which requires waiting until the last split-second to find out if you will be around for the next second.

We saw only what the Russians wanted us to and judged their capabilities accordingly. Looking at their efforts through a tunnel, it was easy to see them as larger than life. NASA and the American public watched their accomplishments jealously and listened to periodic rumors of program setbacks and cosmonauts that were supposed to have died in orbit. I have never seen a single piece of hard evidence to support any claims of Russian space disasters, with the exceptions of those which have been discussed publicly. In retrospect these rumors were little more than wishful thinking. We looked anxiously for signs of their mortality in this contest and breathed a little sigh of relief in the mid-sixties, when we seemed to be holding our own.

Their hardware was simple, as were their early missions. In the absence of first-hand information, we judged the technical excellence of their program by the firsts they were accumulating and the growing capability which we presumed it represented. We felt they were in a race with us to land a man on the moon. If they ever were, I believe they abandoned the contest by 1967 or 1968. At that time they were in the midst of the long interruption of their manned program as a result of the Soyuz 1 accident. Our program was proceeding more smoothly than we had ever expected. Our systems and operational techniques, which may have been overdone in the beginning, were starting to pay off. The Russians, too, were reaping the harvest of their early programs —and that was their problem. The simplified missions and hardware may have been adequate for propaganda victories, but as the game became more technologically challenging they were woefully inadequate.

Analyzing the past fifteen years, it becomes obvious that the Soviet achievements in space have been primarily of the propaganda and public relations variety—not technical. The Apollo-Soyuz mission of 1975 was tailor-made for that role. Who was most in need of space "cooperation" in the early seventies, and also best prepared to exploit it?

How I feel about that flight and the conclusions I draw from it are colored by my experience as an astronaut and have nothing to do with politics. But, frankly, I did not and do not trust the Russians or their space establishment. I would no sooner trade secrets with them than I would with Red China. The last anyone heard, Red China did not have a space program. But who knows? Any country of 800 million people that claims Ping-Pong is its favorite indoor sport would lie about other things, too.

I am less concerned with what we may have given away during the Apollo-Soyuz effort than I am with the risks that lie ahead if this partnership continues. We were, in this case, exposing the Russians to American technology that was ten years old—a moral victory?

Today it is believed that the Russians budget at least $6 billion a year on their space effort. Keeping in mind that their gross national product is half of ours, this means their support is equivalent to three to four times what NASA receives. The Russians are outspending us in space, and what they cannot buy we have been giving them for free. That, in part, was what detente accomplished.

A larger slice of their space dollar goes into an unmanned planetary program, which will certainly leave them better prepared when the time comes for manned landings on other planets. Sound a bit far out? Believe me, it is only a matter of time. Had we made the commitment in 1970, we probably could have flown men around Mars by 1984. As it now stands, with the depressed state of NASA funding, it will probably be the year 2000 before that feat is accomplished.

A little background is in order: During the sixties, the United States frequently took the initiative in urging cooperation in space and suggested joint efforts in a variety of missions. The Soviet Union remained silent. By the early seventies, however, the view of the space world through their telescopes had changed. The United States was landing men on the moon while their own Soyuz and Salyut programs were having problems. To Moscow, it must have seemed an opportune time to hedge a bit. The Soviets have always been opportunists and the U.S. was offering them a chance to regain lost prestige with a minimum of risk. The Apollo-Soyuz mission in mid-1975 was tailored precisely to their pattern.

The planning for the joint mission pushed ahead relentlessly in spite of the fact that our increasing exposure to their program convinced us that the Russians were well behind the United States in both hardware and attitude. Although we expected their equipment to be a Salyut space station, we settled for a Soyuz. When we asked for higher orbits and a longer flight time, we settled for lower and shorter because the Soyuz could not meet the requirements. And while the long-closed doors of the Russian space program opened a crack, they never allowed our training on their own field to approach what was available to them in this country.

By the end of the 1960s the world had declared NASA a clear winner; by 1975 our politicians had deliberately reduced the space race to a dead heat. To those who saw it as a means of keeping our launch schedule active for an additional two years, it was called progress.

It was an odd kind of progress, achieved by reaching down and lifting the also-ran up to our level. The sense of urgency that we felt in the late fifties, when Sputnik sent its chilling signals back to earth causing us to revamp our entire educational system, was now being felt by the Russians. From its inception, space achievement had been viewed as a measure of a nation's scientific, military, and even cultural strength. By that, or any other yardstick, the United States was riding high.

Nevertheless, in the early seventies the joint mission became very important to NASA. It would be the only manned flight between the end of the Skylab program, in 1973, and the first launch of the shuttle program in 1980—or later. When the mission was announced the reaction in the Astronaut Office was like that of a college football team invited to a bowl game. The virtues of the project were defended as sincerely as any of us would have defended additional lunar landings.

NASA selected a crew of real old-timers (average age, forty-six years) although only one had ever flown in space. Slayton, at fifty-one, had been director of Flight Crew Operations for ten years, one of a handful of lieutenants reporting directly to Dr. Robert Gilruth, the director of the Johnson Space Center. His career had suffered that bitter setback in 1962, when he failed a physical two months before he was to lift off as the second of the Mercury astronauts to go into earth orbit. A squiggle was found in his electrocardiogram and it was enough to ground him for ten years. A bug on physical fitness, he ran two miles a day. Not that physical condition was ever a factor in crew selection, but it would certainly be looked at if anyone ever questioned the implications of Deke's age.

It must have eaten at Deke's insides to need another qualified aviator along when he flew. I sat in the other seat on some of those flights, and though Deke was never able to log as much stick time as the rest of us, he remained a superb aviator.

For ten years he had cut himself off from the intimacy the rest of us could enjoy. He operated in a lonely role, seldom sharing his thoughts. We could be working regularly with Deke and worrying daily about crew assignments while he kept them confidential for months after they were settled in his mind.

Deke never failed to back each and every one of us to the hilt, even when we put ourselves in indefensible positions. When our crew carried several items into orbit that did not meet the current restrictions, Wally Schirra was careful to tell Deke about it the night before our launch. We knew we could count on Deke. Unfortunately, Deke's faith in us wasn't always as well justified.

With the Apollo 15 envelope scandal his boys had put him out on a limb once too often. Enforcement of NASA restrictions on what a flight crew could carry on a mission as personal souvenirs had been delegated to Deke by assuring management he would not approve items in bad taste or which violated safety codes. When the scandal broke Deke was in the embarrassing position of leaving the agency open to criticism at a time when it was becoming increasingly difficult to generate financial support in Congress. The Apollo 15 incident and other flight crew peccadillos must have cost him dearly in credibility with his superiors. We simply didn't measure up to the support Deke gave us.

Deke never let up in his efforts to return to flight status, but it turned into a question of time. Could he beat the clock? Could he get ungrounded while there was still a flight left, or before he started collecting Social Security?

Over the years he frequently sat in on our training sessions and most of us got the impression he expected to use it some day. We thought at times it was a little sad, like watching an old actor hanging around Central Casting, hoping for a call. We admired him for it, but in a corner of our hearts we hurt for him, too. It seemed so pointless.

For all his persistence, his constant attempt to keep up on the training program, it looked as though time had run out. But then in 1972 Richard Nixon flew to Moscow and began the chain of events that led to a dramatic space venture between the onetime Cold War rivals.

Slayton, with fifteen years at NASA, was clearly not your typical rookie. Ungrounded in 1972, he recommended himself for assignment to the next flight crew available (which was the ASTP), resigned as deputy director of the Manned Spacecraft Center to devote all his time to training, and began to lobby aggressively for the mission. One might ask, "If Deke was qualified, why shouldn't he get the mission?" The point is, if the assignment was made on the basis of qualifications, or if there were so many missions and so many pilots that it made no difference, then there was no reason why Deke should not make the flight. What the hell, past crew assignments had not always been made on the basis of qualifications, either. Deke had never claimed they were made objectively or on the basis of measurable abilities.

Sentimentally, it could hardly be beat, but as a practical matter, it was another power play accomplished in complete disregard of any objective reasons for doing it differently. There was precedence for playing it the way he did. In 1963 the Mercury astronauts sat down and divided up the Gemini missions such that Shepard got the first one (although he was grounded before it could be flown), Schirra staked out the first rendezvous, Cooper was left with the first "long-duration mission," and Carpenter was frozen out entirely. A more obvious example was Shepard's move from restricted flying status directly to the Apollo 14 lunar landing mission in 1970.

Slayton actually had two campaigns to wage, and he ended up with a 50 percent batting average. Being named to the flight crew was only half the battle. Naturally he wanted to be in command as well. That left Slayton and Stafford on different sides of the issue. Stafford let it be known that he considered himself, by virtue of experience, the only one on the crew qualified for the command responsibility. That had to create some tension between them since Deke was directly responsible for Stafford acquiring that experience. Somehow the case was sold and Deke agreed to take the second seat.

Stafford was Deke's assistant at the time he decided he wanted to fly "the Russian mission." Because he had paid attention to his politics over the years, it really wasn't much of a problem to tie down the crew assignment, but it took a little more work to command it. From Tom's point of view, it must have had everything. Technically it was a breeze. The training grind would be long but repetitive, and Tom could take that in stride. It would be heavy on the social aspects, with a lot of overseas travel and international notoriety and acclaim far beyond the demands of the mission. It was certain to receive massive news coverage. It had, in fact, everything except a real reason for being.

Stafford and Slayton were soon joined on the ASTP crew by Vance Brand. Third spot had opened up when Jack Swigert was ruled out in the aftermath of the stamp-signing-for-pay incident. Vance was a former marine pilot with a boyishly innocent face. For years he slogged away in the trenches backing up several Apollo crews and Skylab before getting his chance. Stafford had been through the training grind so many times he had it memorized and, besides, he was more interested in the peripheral aspects—the travel, diplomacy, new social and business contacts. Deke's spirit was undoubtedly willing but, let's face it, that kind of training is a younger man's game. But it worked out fine. Someone on the crew has to stay on top of everything and the rookie, motivated, dedicated, willing to overkill, feels compelled to do just that. Not surprisingly, Vance Brand was trained to a sharper peak on the technical aspects than his two older crewmates.

The Soviets, for their part, threw in a first-class team for their half of the international space rendezvous. Both were veterans of earlier space missions. Valery Kubasov, the forty-year-old flight engineer, had been a cosmonaut since 1966 and was described as shy but a brilliant engineer. During the joint training sessions, he seldom spoke. In the shadow of his more flamboyant fellow crewman and mission commander, Aleksei Leonov, it was often hard to detect when Kubasov was around.

Leonov was a bona fide Soviet hero. He was one of the original Soviet cosmonauts selected in 1960, and he made an indelible impression on the world in 1965 when he became the first man to "walk in space." Friendly and outgoing, Leonov's behavior contradicted the preconceptions most of us had about the Communist party "organization man." A skilled artist, he made numerous sketches on his mission—including some with religious themes. He has the quick wit and the roly-poly look of a fellow you expect to find emceeing a Rotary Club banquet.

The U.S. crew was training at Star City, just outside of Moscow, when President Nixon paid his historic visit there in June 1974. Naturally, he and Premier Leonid Brezhnev dropped by to shake a few hands. A fascinated Alan Bean, our backup crew commander, watched Leonov in action. "On that occasion," said Bean, "Leonov was one of the first to shake hands with the two heads of state, at which time he engaged them in animated conversation. Standing between the two, he turned around to become a part of the receiving line, and proceeded to direct the remainder of the ceremony."

An air force colonel, the son of a miner, Leonov was an excellent politician, impressed with our technology but always diplomatically refusing to compare the Soviet and U.S. systems.

These five men came together to perform one of the most ballyhooed missions of all time. It was part history and part science fiction, and it created a showcase for several activities which were commonplace in our program but breakthroughs for the Soviets. Given the usual Russian passion for secrecy, it did have the effect of letting in some light. For the first time, the Soviet Union announced the names of the flight crew prior to a mission, as well as the date. For the first time, U.S. observers were permitted an inside look at parts of the Russian space program and the U.S. ambassador was allowed to watch the Soyuz lift-off from the once-secret launch site at Tyuratam, 1,800 miles east of Moscow.

Perhaps even more noteworthy, for the first time Soviet citizens were able to watch, along with the rest of the world, a live television broadcast of the lift-off and landing, scenes Americans had come to take for granted.

Amid all that international fellowship, and the growing excitement of a space spectacular that both joined and pitted national pride, it became more and more difficult to pin down exactly what the objectives of the mission were, but along the way these points were emphasized.

The ASTP mission was described as essential for testing the technical solutions for compatible docking systems between United States and Soviet spacecraft. That is certainly worthy of a high priority for the USSR. Historically, docking has given the Russians more trouble than any other phase of their operations with the possible exception of reentry. Many of their missions have been judged a failure by U.S. experts because of an unsuccessful docking, while we have a near-perfect record in that important phase of space operations.

They had obviously suffered hardware problems, but on this mission both sides were utilizing the same design, which differed only slightly in engineering detail and hardware implementation. That design evolved from a concept proposed by Rockwell International for the Apollo Command Module in the early sixties. The system was put on the shelf at that time because it was much heavier than the probe and drogue concept which we eventually employed. Now we handed it over to the Soviets.

At its conception, the Apollo-Soyuz mission was touted as a "demonstration of space rescue capability." Who could argue with that?

With the space shuttle, scheduled for flight throughout the 1980s, we could be called upon to rescue future space crews—ours *and* theirs. Some question exists, however, if the other half of that commitment could be called into action should the need arise. From their launch site in Tyuratam it will take a much greater capability than the Soviets have shown to date to reach the orbits of most of our manned space-craft. This is no big loss, since with the completion of the space shuttle we will always have our own rescue capability. But it ought to be un-derstood that no *quid pro quo* exists here. In the meantime, for the five years between ASTP and the space shuttle, we will have *no* crews of our own to rescue and *no* hardware on the pad to provide rescue for the Russians.

The only mutually essential element of rescue which we could not verify before the mission remains unproven. The Russians have not demonstrated the operational know-how to rendezvous with one of our orbiting spacecraft in our *usual* orbits. This could be of great concern should we prove to be the rescuee, instead of the rescuer.

In the last sixty days before the launch, more and more knowledge-able Americans were facing up to the fact that they were buying an enormously expensive public relations stunt. In an effort to nudge along detente, the United States government encouraged the three major television networks to pool their resources to bring the country a record total of more than thirty hours of live coverage. It was, in short, more political than scientific.

This wasn't difficult to see. Science on the ASTP was not only sec-ondary, but second rate. Some experiments were little more than a reclassification of earlier, routine data collection, some trivial and some ridiculous. Special consideration was given to experiments that would take advantage of the presence of two manned vehicles in orbit to-gether. This led to such gems as "exchange of microbial growth," which involved trading objects from one spacecraft to the other. Then there was the artificial solar eclipse, achieved as Apollo pulled away from the docking, blotting out the view of the sun from Soyuz. This enabled the cosmonauts to photograph the corona for Soviet astronomers. We ac-cepted this bit of trivia in spite of the fact that our last mission on Skylab had produced hundreds of hours of the highest quality solar coronagraphs.

In the end, the Soviets could not resist a bit of last-minute games-manship. Their press kit described six experiments our planners hadn't even heard about!

It was definitely a mission looking for a reason to happen and, opera-tionally, amounted to little more than a rerun of the Gemini 8 flight carried out by the United States in 1966.

It had one redeeming feature for this side of the world. Though we didn't unlock the mysteries of darkest Russia, we did manage our first real exposure to the Soviet space program, the only other manned space effort in our world. The door opened a crack in 1972, and over the next three years our scientists, technicians, and astronauts gradually and cautiously gained some trust and peeked further inside. The more we saw, the more it confirmed our early impressions. The conclusions were hard to swallow because they were incompatible with the giant that NASA, the western world, and most Americans pictured as the Russian program of the sixties.

Knowing the swift start they had made and the huge lead they once had, we were skeptical when we finally saw their hardware. At first we suspected we were not seeing the real thing. It was almost easier to let our imaginations run wild than to believe the evidence before our eyes. We had overrated the Russian program, their hardware, and their operations. We had seen shadows on the wall and imagined monsters. The fact was the Russians deserved great credit for what they had achieved. They had made the most of a limited capability and converted it into excellent press in the world media, and they accomplished it with hardware that was archaic by our standards.

During the joint training phase the Soviet and American crews alternated three-week sessions every three months at each other's facilities. At the Russian end, most of this time was spent in Star City, a Moscow suburb which is roughly equivalent to the Johnson Space Center at Clear Lake, a suburb of Houston. When Stafford insisted that the mission could not be flown unless they were allowed to inspect the flight hardware, the Russians permitted a one-day visit to the launch complex at Tyuratam.

At Star City the days were spent in loosely structured briefing sessions on the Soyuz spacecraft and the mission. Evenings were frequently occupied by semiofficial social gatherings.

It was those "group gropes" on the vodka circuit with the cosmonauts that really took their toll. At dinner and drinking parties vodka flowed like mineral water. At the risk of losing their macho image, our guys freely admitted that the cosmonauts buried them in the vodka consumption. Bob Overmyer reported he was worried about getting a bill for dead potted plants, because of the great quantities of the stuff he dumped in the nearest receptacle.

Our lads awoke each morning with a new awareness of what Dr. Schweitzer once called "the fellowship of pain." Their Russian hosts took it in stride. The U.S. contingent finally met the challenge by appointing a "duty drunk" for every official or semiofficial party. His job was to keep up glass for glass with the cosmonauts and leave the others free to circulate and survive. The rumor was that Stafford became the sacrificial lamb more than his fair share of the time.

Living on the Russian economy presented a problem of a different sort. Trying to obtain a satisfying breakfast or a light lunch was all but impossible, and dinner always came in multiple courses. Some thought the food at Star City was good, but most would have preferred to eat breakfast at the United States Embassy snack bar every morning. That was impossible, due to the large number of people in the party, and visits were restricted to once a week. That snack bar was voted the best eating place in Moscow.

On one of the training trips in 1974, several of the astronauts were joined by their wives, a rare chance to mix a vacation with business. While the flight crews underwent training during the day, the wives enjoyed the hospitality of the cosmonauts' wives, Svetlana Leonov and Ludmilla Kubasov, and others. Some of the American wives were introduced to the modest touches of status available in a classless society when they were taken on shopping trips in Moscow in their hosts' chauffeured limousines. That by itself was impressive, but when it was invariably accompanied by screaming sirens and motorists pulling off to the side of the road to let them pass, it began to get a bit ridiculous. The cosmonauts' wives seemed to take it all in stride.

All things considered, those visits hit new highs and lows in the training regimen: the most travel ever and certainly the least opportunity for anyone to fraternize.

The American portion of the training took place at the Johnson Space Center. From there, the group branched out all over the country on technical and, just as frequently, social missions. These side excursions ranged from Disneyland in Anaheim, California, to the launch control center at Cape Kennedy.

Compared to past missions, the joint training sessions in the United States must be regarded as nothing less than bizarre. With one of the principal objectives clearly to pump up diplomatic relations, a major portion of the three-week visits were taken up with tourism and public relations activities. For the astronauts, accustomed to eighteen-hour training days, it made for a much more relaxing pace than they had known before. Social activities were usually high visibility. In terms of generating political propaganda, it may have been a legitimate way to "train." But it made a mockery of the preparation which went into earlier missions.

A fair appraisal would be that adequate and appropriate training was accomplished because the technical interaction between the crews —and the vehicles in orbit—was trivial. The real relationship, even during the mission, was social—that handshake in the sky. The theme song for the training period should have been "Getting to Know You."

A good deal of just that took place during the September 1974 visit. The Russian party included their prime crew, *three* backup crews,

General Shatalov (the Deke Slayton of their group), and their interpreters. Just before their departure for home, the combined Apollo-
Soyuz crews held a dinner for the other astronauts and their wives.
Interpreters were in attendance, but it was obvious that the Russians
had a much better command of our language than we did of theirs.
The standard Russian joke was that the mission had three official languages: Russian, English, and Stafford's Oklahomski.

After Stafford provided a few words of friendship on behalf of the
astronauts, he turned the show over to Leonov, who speaks fluently in
only slightly broken English. He has a real gift for turning a phrase.
When asked by a newsman about the merits of Soviet and American
space food, his reply was, "It is not what you eat but with whom you eat
that is important." (During the flight, however, Stafford bolted down
three tubes of soggy borscht and required three Lomotil pills later to
settle his stomach.) Leonov delivered a humorous, impromptu speech
that ranged from the progress of their joint training to jokes about his
success at a recent antelope hunt in Wyoming. He was impressive
without seeming to try.

The other cosmonauts were less at ease with English, but all seemed
to carry a fair dose of political savvy. It was generally felt by our crew
that each of the Russians had additional "obligations" on their visits to
the United States. As Al Bean put it, "The Soviet Union does not have
so many human resources that it can afford to send such capable individuals to this country without asking each to do some additional re-
connaisance."

The personality of Leonov, and his success at charming his American
hosts, pointed up one more area in which the Russians may have
outfoxed us. In Stafford, Slayton, and Brand we had selected good
technicians to do a technical job. All were attractive men and skilled
professionals, but no one would describe them as varsity lettermen in
the tea-and-crumpets league. Their Russian counterparts had either
been sent to a finishing school or been selected on the basis of their
charm, facility with our language, and political instinct. They did a
splendid job. For us to reciprocate in kind, on the same level of congeniality and social footwork, we would have had to send in our first
team in that category: Gene Cernan, Wally Schirra, and Jim Lovell.

With only one simple piece of new hardware, the joint training
concentrated on such things as how to pass emergency instructions
between crews in Russian, dreaming up appropriate remarks to make
in each other's language while the world watched or listened, and how
to train the cosmonauts even a little when at least half their time in the
United States was devoted to public relations activities.

Historically, cosmonauts have primarily been subjects in an otherwise automated mission. This alone accounted for their ability to send passengers into space with only minimal pilot qualifications. Dogs, monkeys, nonflying engineers, and Valentina Tereshkova all survived nicely in the same kind of spacecraft. As test subjects, you would expect their training to be influenced greatly by the doctors—and it is. In addition to usual flight preparations, the Russian medical scientists have control over the cosmonauts for one week in every quarter plus two additional weeks during the year, for controlled medical and physical conditioning. That same authority enforces a six-week vacation leave each year. So for three months out of every twelve, each man is unavailable, and that would represent an unacceptable reduction of training time in the U.S. program.

What the astronauts had suspected on their first visit to the Soviet Union was confirmed on repeated training sessions over the next three years. In the Soviet program, "man in the loop"—meaning man with an essential job to perform—was more theory than practice. Their own requirement to launch and recover manned spacecraft with the entire crew incapacitated resulted in a very specific design for the control functions: a design in which the cosmonaut was little more than a short circuit for a telemetry control signal from the control center. This method of operation has been an integral part of the Soviet program since the first time man was sent along for the ride. A crewman is able to perform only a limited number of the control functions. Man and his infinitely variable computer have not been fully integrated into the Soviet space program. This shortsightedness has severely limited the accomplishments of their manned program and turned their lead of the early sixties into a trailing position by the end of the decade.

Even the Soviets' nominal use of man had enabled them to exploit one of his more important advantages: on-the-spot presence for essential operations. In their early manned shots, before they began deploying tracking ships to the Indian Ocean, ground control was in contact with their spacecraft only while it was over the Soviet land mass. This generally limited contact to portions of the first six orbits and then no contact again until the sixteenth or seventeenth orbit. The cosmonaut in the spacecraft was available to perform necessary functions even when the spacecraft was out of contact with the ground, a very significant portion of the time even in the current Soviet operation.

Russian manned spacecraft are grossly deficient in many areas. There are no on-board meters for continuous monitoring of many essential systems, such as a good attitude reference display. Russian spacecraft have no computer. The pilot cannot inject himself into or improvise unscheduled operations. Since they have no reentry guidance system, a landing spot can only be selected within tens of miles. They have made the most of it, however, and even conducted some efficient recovery operations after the spacecraft has landed. Judging by the deficiencies, one would think the Russian criterion for a man-machine relationship was *simplicity*, while the United States has always emphasized *capability* and *flexibility*.

It is safe to assume that the Russian interpretation of fail-safe is considerably different from our own. Their reliance on automatic or ground-controlled systems demands that failures leave them in the automatic mode of operation. This might have been significant in the fatal Soyuz 11 depressurization during reentry. The pressure relief system included a manual and an automatic valve in series. Procedures called for opening the manual valve prior to separation in orbit to insure that the automatic system was free to equalize pressure on re-entry. The automatic valve cracked open at separation, evacuating the spacecraft and killing the three-man crew.

A similar system in an American spacecraft would require the manual valve to remain closed until the automatic valve had failed to perform. More important, an American spacecraft would not have been allowed to fly when a *single* failure in such a critical system could kill the crew. Equipment and procedures such as these have led me to place their technical level of their hardware between our Mercury and Gemini programs, circa 1965.

One logical explanation for their great reluctance to have a free exchange of information is embarrassment at their own state-of-the-art. Unfortunately, the predominately one-way exchange can better be attributed to the Soviet system itself. Even the simplest technical details are frequently treated as state secrets. One day Astronaut Bob Overmyer spent forty-five minutes trying to get a Russian engineer to explain how their TV camera mounting bracket worked. On the other hand, the U.S. engineers and technicians would go into the most minute detail of our more sophisticated equipment at the drop of a hat.

The most essential requirement of managing a program as complex as the space business is good communications. Here again the Russians are still bogged down in secrecy, and not to keep the information from foreign nationals. There is minimal exchange among managers on a horizontal level and only a small amount up-and-down. Only at the very top levels of management is it possible to encounter anyone with knowledge of more than one system of the spacecraft. Information provided to one working group of their experts never seemed to get passed on to the engineers who attended the next joint working session. Data furnished to one engineer in an organization would seldom be made available to another engineer in that same organization. This was consistent with the trust-no-one syndrome and provided increased job security. To give an example of this, Deke related what happened on one cosmonaut visit to Houston. After some discussion with Leonov and Shatalov, it was agreed that the cosmonaut team would fly to the West Coast at the end of the week. Late on the night before the trip, Deke learned that the travelers had not yet been informed that early the following morning they would leave for the Rockwell Plant 1,500 miles away.

Even the Soviet leaders are not made aware of (or do not understand) the relative strengths and weaknesses in their own space program. This lack of understanding at least *contributed* to the Soyuz 11 deaths in 1971. The story we heard was that in the mid-sixties Soviet leaders, aware that the United States would be flying a three-man spacecraft in the near future, ordered the spacecraft constructor to launch three men in their own Soyuz vehicle. It was explained by their scientists that there was neither room nor weight capability to launch three men into orbit; however, in order to accomplish his directive, he eventually did design three launch positions into the vehicle with the third man relegated to a seat which could best be described as sitting on the shoulders of the other two. In order to place the heavier spacecraft into orbit, the three space suits were deleted leaving no backup in the event of a cabin depressurization—which did occur.

Flight controllers in the Soviet mission control center work with spacecraft data that is at least several minutes old, and data relayed from tracking ships is two to three hours old before it is available for display. By contrast, the Russians were in the embarrassing position of displaying NASA-relayed data from an American spacecraft flying over the Soviet Union several minutes prior to receiving it from their own spacecraft directly overhead.

The conclusion is obvious. After assessing their capabilities we nevertheless pushed ahead with a joint venture in which we had nothing to gain technically and everything to lose. The exercise had to be justified on the basis of political benefits, which proved to be tenuous and highly controversial. Detente might offer relief in some areas, but in space it translated to wishful thinking and a give-away policy.

It is generally assumed that the astronauts have inside information on the Russian space program. Unfortunately, we do not. Frequently we had to turn to such publications as *Aviation Week* as our best source. In truth, we spent little time worrying about what they were doing. We were too wrapped up in our own work to show more than a passing, professional interest in the other space company. It involved us only to the extent that NASA, as a Pavlovian reflex, would pitch for more public support as a result of anything the Russians attempted. In retrospect, a more fundamental belief in what we were doing, and greater confidence in our own approach, would have led to a more objective assessment of our competitor.

Some satisfaction can be derived from knowing we have overestimated the Russian capabilities for the past fifteen years. That awareness could have application to our estimate of the Russian military capability as well. Our new knowledge of the Russian hardware and techniques has convinced me that many of the challenges in space could have been addressed with much simpler systems than the United States committed itself to in the early sixties.

The average age of the crew may have made the *Guinness Book of Records*, and that might have taught us something as well. Flying a spacecraft is really not all that difficult or demanding. There is certainly no obvious reason why a healthy, well-trained, forty-five- or fifty-year-old pilot cannot safely handle it. The training grind is a different story! It is a hectic, physically and mentally demanding pace for anyone, and especially trying on forty- and fifty-year-olds.

It was not surprising to hear that Deke was having trouble during training. He was experiencing first-hand the physical and mental demands of preparing for a space mission. Technically, ASTP was one of our easiest missions and, therefore, a good way for him to break in. Three years to get ready was a far cry from having six months to train for a lunar landing. No one ever gave a better effort but it is the Kenny Stablers, not the George Blandas, that get you to the Super Bowl.

Stafford? Well, he certainly qualified as an old pro, but that can also lead to a casual approach to the job. You can bet he did not pursue the ASTP assignment from an all-consuming desire to go into space for the fourth time. It was the glamor that brought the firehorse back to the fire. Add to that an appreciation for Deke's difficulties, and we have two of the three crewmen taking less than a full measure from the training routine.

The load fell on Brand's shoulders. He was a forty-four-year-old rookie, but motivated as we all are on that first flight. For whatever reasons, the crew did far less training *together* than usual, even for those phases of the mission which require close coordination. Add to this the heavy social schedule, the great emphasis placed on public relations, and the political objectives of the mission, and NASA may have launched this last Apollo spacecraft crew with a minimum proficiency.

The docking of the Apollo and Soyuz vehicles went smoothly, 140 miles above the earth, and they remained joined for forty-four hours. The crews exchanged handshakes and gifts, ate together, spoke each other's language, and smiled into the television camera. When it ended, the only major mistake—a nearly fatal one—was ours, and most of the benefits belonged to the Russians.

The mission was an artistic and political success, but technically the performance was no better than the training. While the handshake in space and speeches on cooperation, brotherhood, and peace were carried out almost flawlessly, "secondary" objectives—such as flying the spacecraft and reentry—were beset with their share of problems. No flights get by without an occasional cockpit error, but the higher incidence of them on this mission, including forgetting the two most critical switches in the cockpit during reentry, could well be a consequence of the crew's casual approach and insufficient training as a team.

For their debut on international television the Russians came through like champs; the landing was perfect. Soyuz touched down on a dusty plain near Kazakhstan, ending a flight that for them—in Leonov's words—had been as "smooth as a peeled egg."

Our own guys had their problems. No one had any idea what danger they were in, including the crew, when a "brownish-yellow gas" began to fill the cockpit as the spacecraft descended from about 20,000 feet. For the next five minutes they found themselves choking, gagging, and fighting for air. After splashdown, they struggled for another five minutes trying to reach the oxygen masks behind their couches. Vance Brand passed out cold but quickly revived when Slayton and Stafford clapped the mask over his face. As the navy frogmen swam toward the capsule, only weak radio contact kept everyone from hearing Stafford yell into his microphone, "Get this f—— hatch open."

Two redundant switches, the ELS (Earth Landing Systems) automatic switches, had not been thrown, preventing the automatic devices from not only deploying the chutes, but also dumping the excess fuel and oxidizer from the attitude-control rocket engines.

If one is serious about staying alive, these two switches are not to be overlooked. Stafford was apparently distracted and neglected to call for them at 30,000 feet, the usual spot. On the other hand, that callout was my responsibility on Apollo 7, and I could have died at 30,000 feet without Wally missing those two beauties. At any rate, Brand missed them which consequently made it necessary to use the manual backup switches. In the resulting chain of events, some of the rocket fuel—a highly corrosive nitrogen oxide gas—was sucked in through the cabin vent valve.

After recovery Stafford, Slayton, and Brand were deposited aboard the carrier and rushed below deck to be examined. It wasn't until seven hours later that any word was released about the seriousness of their difficulty. The effect of such a toxic mixture on the lungs could be permanent and lethal.

As the rookie in the crew, and the Command Module pilot flying the reentry, Brand made a point of shouldering the blame for the failure to throw the switches. Stafford, glossing over the confusion at 30,000 feet, took some of the pressure off by declaring, "As the commander, I am responsible for whatever goes wrong in the spacecraft." The claim wasn't made too strenuously and it was still Brand who had to field the questions and explain the goof wherever he went in the weeks that followed.

Our operational performance left a little to be desired. Coupled with some of the Skylab experiences, it triggered a NASA review of crew functions in consideration of adopting a higher degree of automation.

From the vantage point of an alumnus, there were some disturbing signs: a more casual approach to training, less reliance on the crew, and a lot of artificial emphasis on ASTP at a time when there should be genuine interest in the space shuttle. It didn't help to see $400 million blown on a space technology give-away while the shuttle budgets were shrinking monthly. Our give-away of ten-year-old technology was a good deal only if it eliminated support for future joint ventures with our space shuttle.

We did seem to develop one new national consensus: In the sixties, the United States had built a hell of a lot bigger boogie men out of Russian technology than was warranted. The Soviets had made the most of a limited capability, and it doesn't help to know their new international prestige was achieved with space hardware that was archaic by our standards.

Our timing was perfect for their purposes. At the end of the sixties the United States was recognized the world over as the unchallenged leader in aerospace technology. By 1975 we were able to reduce the situation to an apparent dead heat through our own diligent efforts. There's still one born every minute.

17

What Have We Learned?

B Y 1975, America was fresh out of manned space flights. What did we really have after fifteen years, billions of tax dollars, and eight deaths besides a mountain of rocks to prove we had been to the moon? Once you wade past the stock answers, what was accomplished? The public should question the necessity of those trips. And they will want better answers than the need for earlier weather warnings or Teflon frying pans or knowing the mineral composition of the moon.

The problem is that question cannot really be answered now or even ten years from now. Current judgment is made even more difficult with the U.S. space effort between programs. Apollo is over but the United States is committed to placing the space shuttle, our next generation of space vehicle, in orbit in 1979.

This cannot occur before 1980 or 1981, leaving a five-year hiatus after the Apollo-Soyuz joint venture of 1975. That period will serve to consolidate the gains made in fifteen years of space flight and allow us to reflect on whether it was worth the price. As for myself, I don't put a price tag on romance and adventure.

The question of whether it was necessary, whether we needed to explore a world beyond our own, can't be answered simply. The space effort means different things to different people. The U.S. space program has been the one with the high visibility profile, the one that made the giant leaps. We are the ones who came from behind to the point where our technology and operations are now five years ahead of the Soviet Union. The world has been able to watch *us* and has been most affected by our efforts. But the tribesman in Africa is expected to have a different reaction than middle-class America. Each must judge the impact on his own life. I don't presume to be impartial in my own judgment, but I have tried to be honest about what I learned, saw, and felt in the years when the United States poked a hole in the fabric of the future.

By any yardstick, my eight years with NASA were rewarding and fulfilling. They were also frustrating and confining. Many of the astros would have known great achievement without a manned space program. But most of us would also have had more difficulty achieving the same professional level or the same recognition in any other career. It is difficult to imagine a better opportunity to attract attention to our capabilities. Being educated and motivated individuals, none of us would have wound up on a street corner selling apples.

Regardless of what any of us could have done if we had never become astronauts, whatever we have accomplished after will somehow be related in all men's minds to that experience. We should accept that. For me it was an opportunity to get my candle out from under a basket. However, I strongly identify with Phyllis George, a CBS sportscaster, when she says, "One of my goals is to reach a point where nobody introduces me as a former Miss America."

The day I stopped my Porsche on that mountain road and listened to Alan Shepard lift off in Mercury 7, the entire nation was moonstruck. By the mid-seventies the American public had grown bored with manned space flight. But that is part of our nation's character; as technology accelerates, so does our capacity to become bored with its accomplishments. Compare the present media reaction with that of twenty years ago when scientists could, and did, constantly warn of space dangers: meteroids, cosmic radiations, severe temperature changes, unfiltered sunlight, and so on. There was a feeling that a mechanical spaceship might survive—but could man?

The science editor for *Life* magazine reported in 1953 the opinions of air force doctors: "If humans want to work in the vacuum outside their spaceships they must do it in solid walled cylinders—if they want to walk on the surface of the moon they will have to do so by means of mechanical or leglike appendages."

Wrote a *Business Week* editor in 1957, ". . . There is a long theoretical gap between making an unmanned satellite circle the earth, and building a spaceship in which man can travel through outer space . . . with a reasonable expectation of living to tell about it."

Just as the Oregon Trail became a highway, the hostile environment of space has been transformed into another byway of man's travels with relative ease. Equipment failures, even in-flight crises, have been overcome quickly and often routinely by the astronauts and the technicians on the ground. By 1972 a rocket trip to the moon had become almost commonplace to the American public.

So what if the public takes it for granted. It wasn't until we began taking airplanes, television, and computers for granted that we began to exploit them for man's needs. It was essential that space flight be removed from the mystery and awe that once surrounded it and placed in the realm of public understanding and acceptance if we were to get the full measure of return from our investment. So, while NASA may be crying over the public's current reaction, I'm not.

It takes the long-range view to put things into perspective. In a burst of enthusiasm President Richard Nixon called the first lunar landing "the greatest achievement since the Creation." That does seem to cover a rather broad sweep. It began as the challenge of the sixties, but even today, most acknowledge it as the premier accomplishment of the twentieth century. And so it was that the American flag, less than 200 years after it was first unfurled on our planet, was planted on a foreign body in the universe.

What is our legacy of the past fifteen years? It has to be more than fireproof material, smaller computers, and powdered energy drinks. From the beginning the space program provided a tremendous catalyst for, as well as a challenge to, education. An entire generation has been educated since Sputnik. They take for granted what those of us over thirty recognize as a giant leap forward in the content of education. When the Russians launched that little electronic ball, we found out, for the first time in our history, that not only were we not better at everything than everyone else, but we were not even as good as some people at some things. We had to play catch-up, and it triggered a near-revolution in education. In the decade of Sputnik, United States spending on education doubled; science, math, and foreign languages were upgraded; and the number of high school graduates doubled.

When Neil Armstrong and Buzz Aldrin planted their boots on the lunar surface they stood, literally, on a new plateau for all mankind. From that plateau each of us can look out on a new and more distant horizon. I sometimes picture knowledge as a huge, spherical balloon. Each new generation has a responsibility to blow up that balloon a little bit more. Most of the time, technology expands in little spurts here and there. In the sixties, this nation blew a breath into that balloon and expanded man's knowledge tremendously in many directions simultaneously.

Many people—astronauts among them—have compared that first trip to the moon with Columbus' voyage to the new world. In that analogy, Columbus has now returned to Spain, and some have been convinced that the world is not flat. But Magellan has yet to set sail for other, uncharted ports—just as we will set sail for the planets.

History will eventually show that Columbus' voyage was insignificant when compared to man's first trip to the moon. Man has been flying for only seventy-five years, and he ventured cautiously into space just fifteen years ago. Until our voyage to the moon, until we escaped our own gravitational pull, we had not really freed ourselves from good old Mother Earth. Yet we freed ourselves from the sea millions of years ago. Perhaps the most apt comparison to leaving the earth and reaching a foreign body in the universe is that time, millions of years ago, when the first single-celled animals moved from the sea to dry land.

In 1969, I addressed an audience consisting for the most part of people older than myself. I was describing in detail the upcoming mission of Apollo 11—just how we were going to land on the moon in three months. Every two or three minutes two elderly ladies seated just in front of me would shake their heads, incredulously. They were not alone in that reaction. Even after the event, many people were convinced that it was a hoax and that the rocket had actually touched down in a remote part of the New Mexico desert.

I have yet to address an audience of young people where heads shook in disbelief. Anyone under twenty-five grew up with the idea and takes it for granted. His imagination has not been stunted, as it will be later. As we grow older, things which seem impossible to accomplish are nothing more than challenges to be overcome by the young. When a whole generation grows up taking for granted something as fantastic as flying to another planet, think what their minds can imagine and the accomplishments they can move on to. Man's vision is limited not by what his eyes can see but by what his mind can imagine. Our achievements in space freed the imaginations of a new generation.

I have stated on many occasions that the most notable difference between my work at NASA and my later experience is that I don't see the same pride and motivation in the business world. Look around you. How many people would you describe as truly dedicated, who believe in what they are doing? At NASA, during the Gemini and Apollo programs, there was a sense of every job's counting, of being a part of a movement. What did that enthusiasm and motivation create? Only the most remarkable piece of equipment ever built by man to be operated by man, the Apollo spacecraft. Most of us grew up with the negative comparison between American assembly-line know-how and "Old World craftsmanship." The Apollo spacecraft represented New

World craftsmanship at work. While the rest of the world looked on, it performed and added a new dimension to that phrase so prominent on the sides of World War II crates and seen again on the hull of Apollo: *Made in USA.*

There was still another benefit—complex, difficult to analyze, and to me at least, beyond value—it kept alive for a while longer the thrill and the essence of adventure. How we needed to be reminded in the sixties that there will always be new horizons. Where we came from is no more a miracle than where we are headed.

As much as anything else, the astronauts were attracted to their calling by the opportunity to use their knowledge. Mankind, like the individual, needs an arena in which to test its muscles. The manned space program provided an arena so large as to be unique in the world we knew. When such arenas are no longer available, those men who feel compelled to explore them will become as extinct as dinosaurs, and the spirit of adventure will also be dead.

That thought was eloquently expressed in a message from the government of Australia, left on the moon, along with those of other free nations, by the crew of Apollo 11. The last phrase touched a chord:

. . . May the high courage and technical genius which made this achievement possible be so used in the future that mankind will live in a universe in which peace, self expression, *and the chance of dangerous adventure are available to all* [italics added].

At its peak, the space program cost $5 billion per year. The Department of Health, Education, and Welfare spends that amount every eight days. But comparisons are pointless and I am not, at least directly, trying to justify that commitment. All I have meant to do was to describe what happened during those fifteen years that saw Americans travel beyond the pull of gravity and tell how it affected me and others—and how it changed our world, if at all.

How simple our world looks when we are hundreds of miles above it—or thousands, even hundreds of thousands—observing great problems as though through the wrong end of a telescope. Seeing the earth as a small dot in the universe, fragile, beautiful, innocent, and pure, is a very subjective experience. Reactions to that vision are personal, often abstract, and frequently political.

A better interpretation might be, not how tiny the earth is, but how it sets in the vastness of space. It is a lovely vision to look down on our planet and see unity and serenity because one is so far away. But is that the real world?

While a student at UCLA, I had a friend who was student teaching.

One morning he blurted out that he was thinking about giving up his plan of becoming a teacher.

That was a shock and I wondered why.

It developed that he had been teaching all week in a high school art class with uniformly disappointing results, even when each student was asked to sketch whatever he could draw best.

"So I discussed it with the regular teacher," he said. Her solution was quick and simple. "Ask them to draw smaller pictures; then you won't see so many mistakes."

That was part of the beauty of looking at the earth in miniature from outer space. You couldn't see the mistakes or the boundaries that kept one person's France from becoming the next man's Germany.

Many astronauts use that small, fragile world description on audiences to create a warm mood but it does not necessarily mean they experienced a revelation in orbit. I know of no astronaut who set out deliberately to discover what was in his soul, but there were some who claimed a new consciousness from looking back at the earth. There were others who found God on the moon. And some found six-figure jobs as captains of industry. None of us got into the program for very complicated reasons. We wanted to fly and that was the best game going. We thought of ourselves, most of us, as test pilots. Whatever we did, whatever was accomplished, however we behaved, was related more to that view of ourselves than any duty or calling we might have felt.

Someone once said, "A man is never so much alive as when he is close to death." And Winston Churchill added, "There is no more exhilarating moment than to be shot at and missed." That was part of it! Race car drivers, mountain climbers, bull fighters, ski racers, and others feel it. Their exhilaration comes not from being reckless, but from taking calculated risks requiring considerable skill and confidence. The astronaut has made test flying and space flight his sport—his risk exercise.

The astro is proficient and confident. His strong ego is taken by many as egotism and it can be intimidating to others unless they themselves have great self-confidence. In certain situations he is not above feeling almost omnipotent—knowing he is not invincible, but enjoying the experience anyway. He has a high need to achieve, to accomplish tasks requiring great skill and effort.

A test pilot's duties regularly require him to fly an aircraft and exercise judgment in conditions of flight which have not previously been investigated. The attendant "danger" is of little concern. He needs little approval or support from others, but to find himself lacking in his pursuit is the most painful thing imaginable. Anything which even temporarily challenges belief in himself merely spurs him on to more concentration and a better effort.

Take the time in 1954 when, as a green marine second lieutenant two months out of flight school, I was being trained as an all-weather fighter pilot in one of the early jet night-fighter aircraft. In those days flying fighter aircraft at night and in bad weather was considered the province of senior officers only. A good night-fighter pilot required, most of all, experience and judgment.

I was fortunate to be in the vanguard of an infusion of younger pilots who were being trained in this so-called "old man's game." We were greeted with skepticism, and I did little to change that reaction one morning around 2:00 A.M.

With about twenty-five hours logged in the airplane and a grizzled master sergeant named Dan George as a radar observer, I was returning to El Toro Marine Air Station from a night patrol. I had that world-by-the-tail feeling. In command of the most expensive fighter airplane the Marine Corps owned and with my two-man crew performing like a well oiled watch, I let down for the final approach and landing. The field was fogged in, not unusual for that time of year, and I was set up for a GCA (ground-controlled approach). I must have been born with a perverse streak in my character because I was pleased at the lousy weather. I enjoyed my new proficiency and welcomed the opportunity to impress the old topkick at my side. As we started down the glide slope, with the ground controller calling out headings and altitudes, I was mentally applauding myself for one of the better instrument approaches in my brief career.

Within seconds it nearly became my last. At about 300 feet and glued to the glide slope, the airplane picked up a rapid right drift, a fact which completely escaped me. It finally came slamming into my brain from the frequent heading corrections fired at me by the GCA controller: "Turn left to a heading of three-three-zero degrees, you have developed a rapid drift. Turn left to three-two-zero! You're drifting way off to the right and passing through minimums [200 feet]. Take a wave off! Take a wave off! When you have leveled off straight ahead, give me your intentions." As the controller's voice moved a half octave higher there at the end, I came to my senses and got a good instrument scan going once again. Between 300 and 200 feet I had slowed or stopped my instrument scan pattern and developed a case of vertigo, the instrument pilot's nemesis. It wasn't my first or my worst case of vertigo. But it was one of the few times I would experience it without realizing it was happening. The result was an inadvertent 30° right bank.

It didn't take a genius to realize that at 200 feet, coming down at 140 mph, in a 30° bank to the right, we were seconds away from becoming a big bonfire in the housing development to our right.

As I leveled my wings and began a climbout straight ahead, three thoughts came flooding immediately one upon the other:

"That was the best instrument approach I ever made . . . and it wasn't good enough to get us in."

"At GCA minimums, two hundred feet, there was no sign of the runway or even a light on the ground."

"I'll have to do even better to make it in on my next pass."

After that experience, needless to say, my next approach was a hell of a lot harder on the old psyche. There was still a touch of vertigo, which I was fighting all the way, and in the meantime other pilots were being diverted from El Toro to land at El Centro, 120 miles away.

This time, as we settled down on the glide slope, it dawned on me that Dan George, sitting next to me, was probably not enjoying the challenge as much as I was. He very much wanted to spend the night in his own bed but no doubt wished his getting there didn't depend on the performance of a brash young lieutenant flying an airplane through the murk.

I assumed it would be necessary to descend below the minimum altitude to make it in, but in those days that was what I thought it meant to be an all-weather fighter pilot. I was concentrating so intently on the glide slope and heading control down low that it came as somewhat of a shock to feel Dan's tapping on my right knee, indicating he could see the runway lights, at about the same time my wheels were touching down.

Only as we taxied off the runway did the adrenalin begin to kick up a bit. As I listened to the next couple of airplanes execute missed approaches, I dared not ask Dan what he thought. That would have been against the code. But my own feelings were of a challenge accepted and met, not of a risk stupidly taken. I was proud and, yes, even a little excited.

That emotion lasted right up to the moment I climbed out of the cockpit to be met by our operations officer. He quickly informed me that I was the only pilot stupid enough to land in the preceding half-hour, and that I should consider myself unofficially grounded until the commanding officer decided if he wished to make it official.

In the end, it all blew over and I was left with a little proud satisfaction, but it still stands as one of the dumber moves I made in twenty years of flying. Yet, that kind of reaction to adverse circumstances has carried me through a career. On one's mind at such a time is not fear of the task, but the fear of failing, of not measuring up to self-imposed standards. Why be apprehensive about getting hurt or even killed when you may be facing the only opportunity to meet those standards? In this business, it isn't uncommon to be less concerned about getting killed than about making an ass of yourself—especially in front of your peers.

The Gemini abort situation is a good example. Two pilots sat side by side, very close, each with an ejection handle between his knees. If either crew member initiated an ejection, both seats would go simultaneously to avoid having the rocket exhaust from one side burn the pilot on the other. During launch, especially the early phases, conditions could deteriorate so rapidly that the difference between a successful and an unsuccessful abort was measured in milliseconds. The reaction to a typical abort condition was a race against the clock.

Against that background, you might picture each crew member with his hands on the handle, poised for instant action during lift-off. They were alert all right, but most of their attention was directed at *not* aborting. It wasn't unusual for *both* pilots to place their hands in their laps to reduce the possibility of an unnecessary and embarrassing abort.

No Gemini crew ever expressed to me concern about their safety during that extremely time-critical launch phase, but nearly all of them, at one time or another, mentioned the "nightmare" of an unnecessary or premature abort and an empty spacecraft being placed successfully in orbit. The risk to their lives was of no great concern when compared to the "nightmare" of performing badly.

Not surprisingly, astronauts are frequently queried about fear, and just as frequently described as courageous. The question never comes from the fighter pilot, the race driver, or the man engaged in his own risk exercise. To someone sitting at a desk, it may seem dangerous and foolhardy to fly in space. To the fighter pilot physically, mentally, and psychologically prepared to make such a trip, it is but a slight extension of his prior experience.

The people I encounter who say they want to travel in space are generally romanticists without an awareness of what the job entails. For them it is a foolhardy, not a calculated, risk. Those who perform well in dangerous professions feel comfortable about what they are doing. They are not romanticists, but cold, hard realists about their profession. I have never discussed fear with another astronaut or with another fighter pilot. Fear is a minimal consideration in their lives.

I believe an astronaut addresses himself to three fundamental challenges from which he derives his love of the job. The first and easiest is man against machine. The Apollo spacecraft has neither heart nor soul, but is perhaps the most complex piece of machinery ever constructed.

The astronaut masters the mechanical creation of man. The better the pilot, the more he is able to get from the machine.

The second, man against man, is a more challenging contest. To reach the enviable position of riding a rocket, we were the product of a tough and lifelong screening. Since we were selected for knowledge, skill, and competitive drive, the flight becomes a place to vindicate that judgment. I suppose in a sense we are saying, "I am a winner; watch my victory lap." Winners become stronger and go on to prevail over other men in even more trying circumstances.

At the highest level of competition we find man competing against himself. It is that pursuit which provides the satisfaction of being "the master of my fate, the captain of my soul."

As a student pilot, at the age of twenty, Charles Lindbergh concluded, "If I could fly for ten years before being killed in a crash, I would be willing to trade an ordinary lifetime for that experience." I have shared that feeling and know that experience.

Any idiot can get an airplane off the ground, but an aviator earns his keep by bringing it back anytime, anywhere, under any circumstances that man and God can dream up. A pilot's only real obstacles are the elements and the problems he creates for himself—in his mind. Bringing a single-place fighter aircraft home at night with low ceilings in fog, rain, or ice escalates the contest. Apprehension about one's ability to handle the elements can create a crisis regardless of piloting skill. For twenty years my greatest personal satisfaction came from flying anywhere, anytime, in any kind of weather. What some would consider poor judgment was to me a way of life.

A flight to a strange airport, with marginal weather and barely adequate fuel, can be one of the most exciting and satisfying experiences in life. When the radio crackles with the news the weather is worse and threatening to close the field, it can cause an anxious moment. In the next instant my nervous system responds appropriately: the heart picks up the pace and a bit of adrenalin is pumped into my system. This is followed by the thrill of a challenge accepted, and sometimes a self-conscious smile. The heavens have thrown down the gauntlet and I am already anticipating the warm glow, even a smug feeling, after landing. Do your damndest, elements. An intelligent man must recognize he may not win someday, and yet that has always been my reaction.

The astronaut peer group is one of the most competitive ever assembled. For some, selection into this group and flying in space satisfied their highest ambitions. For most it meant only acceptance into an elite fraternity from which to accomplish many of the other things they wanted out of life. But for all of them, from the minute they were accepted, it triggered a renewed effort to rise to the new top among the best in their chosen profession.

An astronaut cannot afford to be awed by his job. When he has developed the confidence, born of proficiency, that he is bigger and better than any portion of the task ahead, then and only then is he capable of performing at his best. The definition of courage I like best is "the capacity to overcome fear." No fear—no courage required. The man who arrives at that special moment in his life, and feels in complete control of those moments where fear might intrude, does not require courage to fly his mission.

I was certainly not courageous. My approach has always been to understand all I can of whatever may affect me, to reduce the unknown elements to a bare minimum. Identifying those moments in the mission when my neck was sticking out, and then knowing when they were over, made it easy to tolerate the brief periods of exposure to physical harm. Years of effort went into minimizing those periods when we had no control over the outcome.

At launch, in theory, the spacecraft has an abort capability setting on the pad. But all crews realize if a pad abort is necessary, the circumstances are conducive to getting killed. Therefore, from some point late in the count till sometime shortly after lift-off, your neck is out. As the spacecraft gains altitude, we gradually obtain more control over the situation and pull our necks back in. A small weight was lifted from my shoulders at lift-off plus one minute when we disabled the automatic abort circuits. It turned into a picnic after two and a half minutes when we jettisoned the launch escape tower.

Sure, if the landing parachutes don't open, it's a bad day. But why spend an eleven-day mission worrying about the chutes? There is no cause for concern until they have a real and immediate influence on your destiny.

It should not be surprising that those first steps (both Russian and American) were taken by a bunch of brash young fighter pilots who approached their missions with anticipation. A good fighter pilot lives every day (only a little self-consciously) thinking he is "the best fighter pilot in the world." It may not be rational, but it is certainly real. For such a man it is only a short step into space.

The fighter-pilot image of wine, women, and song is one all of us have cultivated. That may be nothing but image because flying is truly our mistress. It is unfair competition for a wife and family at times. When I strap on a high-performance airplane, I feel more relaxed, more alive, more filled with self-esteem and confidence than I do at any other time or place in my life. Doing what a fighter pilot does— alone—requires great maturity, but at the same time the fighter pilot is the child that lives in all of us. Airplanes are his toys, and it's tough to give them up. Lets face it, after all the high-blown reasons are torn apart, we remain fighter pilots and test pilots because it's fun. That lifelong fascination with airplanes can easily become an obsession. Giving them up cold turkey is a tough decision that some just will not face.

William A. Anders, who was confirmed as the U.S. ambassador to Norway in 1976, has been a fighter pilot since the early 1950s and has never even tried to take the cure. After flying the Apollo 8 mission in 1969, Anders became executive secretary to the National Space Council, then a member of the Atomic Energy Commission, and later the first head of the U.S. Nuclear Regulatory Agency.

During those six years, Bill did his usual thing: applying himself to his job from dawn until well after dark, which left him little time for flying. Most would have given it up, but Bill worked out his own accommodation. He accepted the Washington appointments on the condition that he remain eligible to fly NASA airplanes. He then used his influence to insure that former astronauts in government jobs were permitted to fly NASA aircraft left in the Washington area while their pilots were visiting on official business. Thus, if Alan Shepard flew a T-38 into a local airport for two days of meetings, it wasn't unusual for two or three former astronauts in the area to fly it before it was returned to Houston. Broken airplanes from this activity have caused their share of arguments.

On many a morning Bill Anders, with a full day ahead of him, would rise at 3:00 A.M., drive to the field, fly a T-38 to some distant field, refuel, return, and walk into his office at 8:00 A.M., ready for work. In late 1976, Ambassador Anders in Oslo was still flying helicopters off of visiting U.S. Naval ships and returning to Houston every couple of months for his jet flying.

That kind of passion for flying, that joy (or craziness), partly explains why a fellow will hang around NASA for ten years, waiting for a space flight that might never come. Is there anything better than cutting through the sky in something new and wild and exploring the unknown?

Why we became astronauts, and how we came through that experience, is a complex question. Not the least of the reasons is that our instincts are those of a fighter pilot. Yes, I was fortunate to take part in "man's greatest adventure," but most of the things I am or ever can be have their roots in an airplane cockpit.

The airplane cockpit and manned spacecraft cabins have now been brought together. For the next quarter of a century, the lion's share of this nation's space payloads will be carried via the space shuttle. Getting the cost of a pound in orbit down from $1,000 to $150 is being accomplished by the development of reusable hardware. The space shuttle was originally conceived as a flyback winged booster and a reusable winged orbiter vehicle. Because of the political and financial climate in the early seventies, along with technical problems, the reusable booster fell by the wayside, leaving the major ingredient of the program the reusable orbiter vehicle. The orbiter is roughly the size of a DC-9 jet, carries a crew of four, and has a large cargo bay and provision for passengers. The payloads of most of the present-day throwaway launch vehicles fit into the sixty-by-fifteen-foot cargo bay— everything from unmanned satellites to the European Space Agency's Space Laboratory.

Once the orbital mission is complete, reentry into the earth's atmosphere is begun as a spacecraft. Below 50,000 feet the orbiter begins to perform as an airplane, or more accurately, a glider, coming to rest on a runway. In less than two weeks it can be ready for another trip.

The shuttle is the most logical and essential step we have ever taken in our space program. However, it not only faces the usual technical risks, but it also has a high potential for further eroding NASA's credibility with Congress and the public. Apparently, tremendous technical achievements cannot be sold to or accepted by the layman based solely on the facts. In the 1960s NASA could not, or perhaps chose not to, sell the Apollo program on the basis of fundamental research, or the development of a new frontier for America, or even on the basis of technological parity with the Soviet Union. Instead national support and funding was tied to the slogan "A man on the moon in this decade." In the long run this public relations catch-phrase understated our achievements and their eventual applications. Our manned landing on the moon was but a comprehensive demonstration of the equipment and techniques we had developed in the preceding ten years. Predictably, everything was anticlimatic after Apollo 11, and support dwindled when that man was actually placed on the moon in 1969.

In the 1970s, using a rationale of the end justifying the means, our only manned space project is being packaged in whatever way is necessary to obtain the funding. The crucial factor this time around is cost. The space shuttle is carried on the books as a $5.5 billion project, and the economics of payload cost reduction are based on one hundred flights per vehicle, sixty to seventy-five launches a year, and a total program of approximately 400 flights. With payloads up to thirty tons, that adds up to thousands of tons a year for many years to come. Since there is no waiting list of commercial payloads, NASA is forced to rationalize the construction of fantastic new structures in space, such as manufacturing facilities, solar power plants, or space colonies. If Congress would not face up to the $6 billion cost of a recoverable booster, is there any reason to think it will be willing to supply the hundreds of billions of dollars necessary for these visionary concepts?

The size-of-the-payload problem is appropriate for a vehicle monstrous in proportion and size. It reflects its breeding—a cross between an airplane and a spacecraft. As an airplane it is an ugly duckling, roughly the size of a DC-9 afflicted with elephantiasis. It is obese from head to foot, with a short stubby wing that is six feet thick at its root. It resembles a slightly sculptured rock and flies with only a little more grace. Instead of a sports car it's a Mack truck entered in a soap box derby.

As a space craft it turns into a swan, with separate flight deck and living accommodations that include a water closet with commode. In the huge rear cargo compartment is a hand-operated manipulator arm that gives its operator the power of a giant. Past crews have been recovered as a lump dragged from the sea after splashdown, much like a bag of cats plucked from a watery grave. The orbiter offers the promise of a smooth landing at the destination airport and the crews stepping from the spacecraft in front of a waiting throng in the dignified manner that a hero ought to.

An example of the unusual solutions devised to meet the challenge of flying this "glider" back from the reaches of space is the thermal protection system. Its airplane appearance does not relieve it from performing one of the most critical spacecraft functions: reentering the earth's atmosphere. Earlier spacecraft have utilized a heat shield to combat temperatures up to 5,000° F. on reentry. The heat was dissipated by melting and vaporizing the heat shield material. It literally "burned" on reentry.

The same solution was not feasible for a winged craft, and a special insulating material was developed. This vitreous material is such a good insulator that it can be held with the bare fingers immediately after removal from a 2,300° F. oven. Much of the surface of the orbiter is covered by bricks of this material literally glued on in what is undoubtedly the most expensive brick laying job in history. Those areas subjected to the most intense heat, such as the leading edges of the wings, will be protected by another new material called reinforced Carbon-Carbon. This solution to the heating problem, as different in appearance as it is unique, is the product of NASA research during the past decade.

On the surface the technical aspects appear to be well in hand, and publicly NASA has shown little concern at being unable to meet both the technical milestones and the schedule. That is also the official position of the contractor, but among the working troops that optimism isn't shared. My own concern is deep-seated.

Will Carter has worked for the space systems division of Rockwell for fifteen years, so his familiarity with manned spacecraft goes back to before the Apollo program. Will had always been enthusiastic about everything Rockwell did. Now, for the first time in my memory, he had a poor attitude which reflected the general morale at the contractor's. In the early seventies the NASA well just about ran dry. For several years their budget was held near its lowest level while salary and housekeeping expenses climbed higher than during the Apollo program. What little money was left over for the contractors produced designs and hardware whose acceptability was sometimes debatable. Systems, hardware, and tests which could be classified as noncritical were pushed as far downstream as possible to be picked up later in the program. The wishful thinking prevails that after "enough" funds have been committed, some of the delayed essentials will be funded to avoid the alternative of an unsatisfactory program.

Experienced and knowledgeable people at the contractor's are aware of the present deficiencies caused by reduced funding and attempts to maintain the schedule. Under the pressures of holding down costs, decisions are made which are frequently in direct conflict with the best interests of the program. There is no more graphic demonstration of the impact these forces can have (with the added pressure of maintaining a schedule) than the disastrous Apollo 1 fire in January 1967. Management in that instance was paying homage to the false gods of cost and schedule in spite of the magnitude of the task at hand. It seems impossible that such an experience could have been forgotten so quickly.

The attitude among the astronauts expected to fly the first orbiters is also different from the old days. Their enthusiasm doesn't come out as, "This is the greatest flying machine since Wilbur and Orville, and I'm lucky as hell to have an opportunity to fly it," but more as, "I've devoted most of my adult life to being an astronaut. If I'm ever going to fly in space, it will have to be in this machine. If that's it, that's what I'll fly." I understand their feelings and would undoubtedly share them if I were still an operating astronaut. They are not unlike the feelings of resignation with which the scheduled flights of the first Apollo spacecraft were approached. We too knew there were problems, and shortcomings, and corners that had been cut. The awakening which placed our "go fever" in its proper perspective was rude and costly: the fire on Pad 34.

The shuttle astronauts come by their attitude with justification. Working on the space shuttle for the past several years, they have had their share of frustrations. Those without a space flight notched on their belt are likely to encounter the "What does he know, he hasn't been in space" syndrome, while those with the magic goofus dust of space on their shoulders are too late. ("We finalized that six months ago.") They have run up against the dollar barrier and fallen victim to the NASA program manager's paternal "We know what's best for you" attitude.

Strangely enough, those who think they know what's best for the flight crew, in spite of what they request, are frequently those who know the least. This is a result of the system in which the on-going program soaks up the best people available in all the disciplines. Those who are left pick up the next program down the line, which may be suffering neglect from top management for the same reason the technical experts are concentrating on the current program.

In that respect there was a lot of similarity between the first several years of the Shuttle and Skylab programs. In 1968 the Skylab management team was headed by Bob Thompson, a NASA survivor since Project Mercury days. It was Thompson's first full program responsibility, and every operational system improvement came grudgingly. The program was a mess. Flight crew input was not received with any enthusiasm until Thompson was replaced by Kenny Kleinknecht in late 1969, and Thompson found a new home, as head of the shuttle program office. On Skylab he was always facing a dollar problem because it was funded on the leftovers from Apollo. With the shuttle, he had no other major programs to compete with, but it was still leftover dollars—leftover from the funds necessary to sustain the NASA organization—and he was still not receptive to flight crew input.

It is tough to assess the impact of another factor. When Skylab was getting off the ground, Dr. Robert Gilruth was still center director, and he was always considered "on the astronaut team." By the seventies he had retired and been replaced by Chris Kraft. While the Astronaut Office and Kraft's former flight operations directorate always looked to each other for support on any operational requirements, we frequently found ourselves at odds when it came to implementation. Couple this with Kraft's longtime resentment of the hero worship accorded the astronauts, and you can appreciate that the line of appeal above Bob Thompson was not particularly helpful.

Astronauts have been assigned to the shuttle program since 1970, but the few "big guns" remaining gave it little attention till 1974. Their usefulness has been limited by the attitudes of the program manager and others on up the line.

From the program manager's point of view a change can frequently be summed up in the weight, schedule, and cost impacts. The flight crew is guided almost exclusively by two factors: crew safety and operational requirements. Program management does not ignore either of those critical items, but their performance is measured against many other criteria as well. For the man who puts his tail in the trap it is a much simpler process. Unfortunately, in the final design and planning phase, when the program office weighed the factors, many decisions went against the flight crew. For example, most of the shuttle crews will have no abort capability for the first part of boost. The drag chute, normally used for high-speed landings, has been deleted. And at one time, the ejection seats planned for the early test flights were scheduled for deletion without it even being discussed with the flight crew.

The impact of these decisions is more than the obvious ones of compromised hardware and operating limitations. The pilots who will fly the orbiter have seen compromise after compromise. One of those pilots told me, "You know, Walt, for the first time in my career I look at the most advanced new flying machine and I'm just not sure whether I want to fly it or not."

That is a hell of a state of affairs!

But they are right. The program isn't in the greatest shape, not only operationally, but also in respect to costs and schedules. Every effort has been made to temporarily eliminate many legitimate program costs that must be added later or permanently allocated to less identifiable budget categories. Another ploy that has virtually been forced upon the agency—by a Congress with its mind on current voter appeal—is the projection of a minimal cost program until sufficient funds have been committed to make cancellation more costly than continuing.

One of the earliest cost estimates for the shuttle program was the 1970 figure of $17 to $18 billion, which included a recoverable spacecraft *and* booster. Repeated direction of "back to the drawing board" finally produced a price tag of $5 billion. The recoverable booster was a casualty along the way.

The shuttle concept of reusable hardware is the most logical step we have ever taken in our manned space effort, even if it eventually costs $12 to $15 billion, which I believe it could. In selling it to Congress on the basis of low estimates, NASA risks its credibility and taxes its management ability when the actual costs eventually go far over budget. The $400-million indulgence in the ASTP technology transfer to the Russians would sure have come in handy on the shuttle program.

While the credibility gap in shuttle costs is undoubtedly motivated by the desire for a rosy picture outside of NASA, the imaginative and optimistic program scheduling is more for the in-house NASA types. For years NASA maintained a schedule of glide-testing the orbiter in 1977 and orbital flights in 1978. One thing after another placed pressure on that schedule, but NASA stood firm. The only scheduled flight milestone that can be met is the first glide test in 1977, with orbital flights not probable before 1980 and possibly as late as 1981.

Under the pressure of schedule and dollars, it is easy to see how operational needs and hardware capability can suffer. The best program is achieved by an honest give-and-take between those who must answer to the budget and those who bring home the vehicle. As a consequence of the past several years, when holding down the cost was paramount, schedule slips will be measured in years, the vehicle will be marginal, and the final cost will be two to three times the announced $5 billion.

It would be a bleak prospect were it not for the fact that these situations do not continue indefinitely. They did not for the ill-fated Apollo 1, although it took a serious accident and the deaths of three good friends to buy the delay necessary for the eventual success of the program. We are not likely to get into such a bind again, but a long delay in the schedule is almost certain between now and the orbital flights. Closer to the "doing," the men who will "do it" carry more and more influence. Not everything will get corrected, but most of the essentials will, and a modified version of the orbiter will be successfully flown.

The pilots and passengers will probably be modified as well. In 1976 and 1977, NASA conducted an eighteen-month solicitation for the best "persons" for the job, with special emphasis on women and other minorities. We can only hope that the yardstick by which they measure candidates is not a flexible one and that those who make it have an overabundance of desire to "dare greatly."

The real shame is that once more we had an opportunity to do it right the first time, and we chose to place our priorities on cost and schedules. Crew members have been working for years to see that a new era of space flight is entered with maximum pilot safety and essential operational equipment. But does anyone at NASA care any more? It's as if the pilots have been yelling into an empty warehouse.

A great deal of credit for the success of earlier programs must be attributed to flight crew contributions to the design of systems and operational planning, the historic test pilot's role. The Original Seven astronauts were tough-minded test pilots who refused to fly a vehicle that didn't incorporate their operational opinions. Bob Gilruth, with Project Mercury, acknowledging the experimental nature of their endeavors, was quick to accept.

The same players filled the same positions in the Gemini program and were equally successful. The flight crew was augmented at that time by an even more capable group of test pilots.

The Apollo program, in retrospect, took a step in a slightly different direction. NASA, expanding tremendously, brought in many new managers. Some were very capable but without the experience of the two previously successful programs, and flight crew members never achieved the level of influence they carried during Mercury and Gemini. With the selection of my group the astronaut corps doubled in size to thirty but, at the same time, the test-pilot experience level was reduced. The hardware contractor was also a brand-new face starting from scratch. These three factors all contributed to the rocky start of a program that was eventually very successful. It achieved its goals, but not as smoothly as Mercury or Gemini, and the flight crew must share the blame. My most disturbing personal recrimination is that we (the astronauts) did not do enough to prevent the fire on Pad 34. Hardware managers completed a program in which three crew members lost their lives, some missions were less than 100 percent successful, and in the end we gave away the store, technically speaking, to the Russians.

Now the country is faced with the most important and innovative step since Alan Shepard's fifteen-minute ride down-range. Flight crew input is a shadow of what it once was, and what little gets through is not met by a receptive management. As Dave Scott puts it, "The contractor thinks he has a spacecraft that can be built like an airplane, when it is really an airplane that needs to be built like a spacecraft."

Some problems should be anticipated, but we don't have to accept the sharing of our shuttle technology for nebulous propaganda benefits. The Russians have expressed a desire to participate in more joint missions when the shuttle begins orbital flights, and NASA has continued with joint working groups. Only time will tell whether we have learned that aerospace technology is an expensive national resource to be jealously protected.

The next few years are faced with many problems, but there are also some encouraging signs. NASA, which since the end of the Apollo program has seemed more intent on perpetuating itself than on pursuing realistic objectives, is changing its perspective. It is no longer looking at space as "a place to go to," space is now "something to do with"—using space to achieve other objectives.

Along with launching the space shuttle in the 1980s we will see unmanned probes to all the planets in our solar system, and maybe even manned visits to some by the year 2000. The last fifteen years have placed these accomplishments well within our grasp.

Man's real limitation in the foreseeable future is the boundary of our solar system. That remains the next barrier to be broken, just as we broke the sound barrier and the earth's gravitational pull.

We must, and will, travel to Beta Centauri, our nearest solar-system neighbor— 4.2 light years away. Will man tolerate spending four years even at the speed of light to make such a trip? Perhaps not. But I do believe that what the mind of man can imagine, he will eventually accomplish. It is only a matter of time before man will further expand his operating universe and take that step to Beta Centauri. We will find a better means of propulsion—if necessary, even finding a way to travel faster than the speed of light.

Our years of achievement will continue and I will glory in them as I did in the sixties. But I remember that Apollo 7, my flight, grew out of an accident, and I wonder what we have learned, after all.

Index